You BETTER WATCH OUT!

TECHNOLOGY, CHANGE, SOCIETY, CULTURE

A. VENKATASUBRAMANIAN

INDIA • SINGAPORE • MALAYSIA

Notion Press

Old No. 38, New No. 6
McNichols Road, Chetpet
Chennai - 600 031

First Published by Notion Press 2019
Title by Sahana Anand
Cover Concept by A.Venkatasubramanian
Content Editing by Vidya Ramakishnan
Edited by Vidya Ramakrishnan & Sujatha Ramakrishan

Copyright 2019 © A. Venkatasubramanian 2019
Red Brick Green Back Consulting
All Rights Reserved.

ISBN 978-1-64678-043-3

This book has been published with all efforts taken to make the material error-free after the consent of the author. However, the author and the publisher do not assume and hereby disclaim any liability to any party for any loss, damage, or disruption caused by errors or omissions, whether such errors or omissions result from negligence, accident, or any other cause.

While every effort has been made to avoid any mistake or omission, this publication is being sold on the condition and understanding that neither the author nor the publishers or printers would be liable in any manner to any person by reason of any mistake or omission in this publication or for any action taken or omitted to be taken or advice rendered or accepted on the basis of this work. For any defect in printing or binding the publishers will be liable only to replace the defective copy by another copy of this work then available.

Dedication

This book is dedicated to the youth.

It has been dedicated to those young and restless souls who are looking to shape and ride the future.

This book is dedicated to the geeks, the misfits, the restive, the crazies and social outcasts who are dissatisfied with the way the world is.

These folks find the prevailing ways of the world limiting, outmoded and cumbersome.

These are the people who want to live life on their own terms.

They want to shape the future and want to shape it in their way.

They are passionate and want to reach for the skies in their pursuit of happiness.

They are the 'change agents' looking to shape a better world for themselves and others.

May they live their vision.

May the future be theirs.

Hats off to them.

Take a bow.

CONTENTS

Foreword — 7

Acknowledgement — 9

Introduction — 11

Chapter 1: The Pace of Change — 15

Chapter 2: How 'Information Exchange' Accelerates the Pace of Technology and Change — 21

Chapter 3: What is your Beef about? — 26

Chapter 4: The Future of Education — 35

Chapter 5: Electric Vehicles — 44

Chapter 6: Automated Flying Transport and Driverless Cars — 51

Chapter 7: Transport for Hire — 57

Chapter 8: Solar and Clean Energy — 61

Chapter 9: On-Demand Manufacturing and 3D Printing — 68

Chapter 10: The Future of Commerce, from E-commerce to K-Commerce — 74

Contents

Chapter 11:	The Future of Healthcare	80
Chapter 12:	Ageing and Disease	90
Chapter 13:	Nanotechnology – Small is beautiful	100
Chapter 14:	Artificial Intelligence	107
Chapter 15:	Quantum Computing	117
Chapter 16:	The Future of Crime and Enforcement	121
Chapter 17:	The Future of Gaming	130
Chapter 18:	The Future of Conflict	139
Chapter 19:	Marriage and Children	152
Chapter 20:	Space and Colonization	161
Chapter 21:	Society and Culture	167
Chapter 22:	Balance of World Powers	174
Chapter 23:	'Creative destruction' at its best	181
Chapter 24:	The Future Czars of Technology	190
Chapter 25:	Future of 'Wealth'	195

Conclusion — *205*
End Note — *207*
About the Author — *209*

FOREWORD

"Human Mind is inquisitive, yet fearful. Conquer the fear by leveraging inquisitiveness."

– Dr. Aloknath De

Throughout history, technological innovation has fuelled human progress. In recent times, the pace of innovation has accelerated—the speed and scale of technological progress has become unprecedented. While handling a wide gamut of emerging technologies in Samsung, I keenly observe that alongside technological innovation comes multidimensional change in our everyday life. Change always brings in challenges to the established order of doing things and therefore resistance to change naturally springs up. Magnitude and velocity of change could only make things worse!

Religion, society, governments and cultural anchorages often resist change as these could stand against the prescribed way of execution. The reasoning, logic or practical benefits of societal change get eclipsed amidst inertia and uncertainty. However, for every technology that could benefit humans, there have been savvy entrepreneurs who have daringly macadamized new avenues forward. Amongst us, there have always been the early adopters and crazy believers who have spearheaded adoption and pushed the envelope of human progress.

Earlier, the generation gap of ensuing age groups was separated by thirty years; with rapid spree of change, generations are now demarcated by mere five years. The rates of adoption and adaptation have increased as generations have gone by. When people do not seek the comfort of the past and embrace change with open mind, they could potentially unlock the future opportunity. Such people understand well that change is the only constant. Many of these trendsetters also, in turn, act as change agents guiding the population-at-large to the benefits of new technologies.

For you, this book 'You Better Watch Out!' could serve as that change agent! The author has delineated how technology is ushering in the society and how it would impact our lives. The chapters in this book often weave historical perspectives with how the future technologies could pan out. The author captures well how the winds of change could even impact culture and sometimes increase friction as the society adapts. The book gives the reader a window into the future and prepares oneself to face the floods of change that are imminent.

The book introduces, for uninitiated, nuances of cutting-edge technologies like Artificial Intelligence, Nanotechnology and Quantum Computing. It not only depicts how society is now embracing Transport for Hire and the concept of Shared Economy, but also shows how it is preparing for Electric Vehicles and Driverless Cars. The author zooms in, with resolution, on the Future of Education, Healthcare and Wealth. The book takes snapshots of themes like Society and Culture, Marriage and Children, Ageing and Disease. The author also strives to deal with complex topics like Balance of World Powers, Space and Colonization.

The human mind is inquisitive. But, at the same time, it is fearful. For most, the best way to conquer fear is to leverage the inquisitiveness. To know how we got here, we need to know the past. The past serves as a reference point for one to see how much life has changed as one travels to future. Future is a series of possibilities, some dreamt of and some even beyond comprehension. The book by A. Venkatasubramanian is an endeavour to open the minds of the readers to what the future could bring. Our lives are going to be in a state of constant flux and one would require to navigate with the times just to stay relevant and possibly an inch ahead. This book helps you do just that!

– **Dr. Aloknath De,**

Fellow-Indian National Academy of Engineering

ACKNOWLEDGEMENT

I would like to thank all those **teachers and professors** who have **piqued my curiosity** over the years and **not cramped my style**. The **very foundation** of **my learning, curiosity and thinking** owes a lot to these **selfless guiding lights** who have **shown the way** with their **love for knowledge** and **love for their students.**

I would like to **thank my daughter** whose **boundless energy and bubbly enthusiasm** has kept me going. Her **child like curiosity** and **sponge like absorption** never ceases to amaze me. She showed me that **human potential is limitless** and is **a living example of that**. I also **thank her** for coming up with the **title for this book.**

I would like to **thank my wife** who has been my **sounding board, content editor** and also **instrumental** in the making of this book. She actively **participated in the discussions and thoughts** that helped evolve this book. I would also like to **thank her and my mother-in law** again for diligently editing this book at various levels.

INTRODUCTION

Where the future and the change it brings, is heralded

"Live as if you were to die tomorrow. Learn as if you were to live forever."

– Mahatma Gandhi

"Future can hit you hard, but just like a wave you can surf it, if you spot it coming at you."

– Author

Most often people fear the future. A majority of people have anxieties related to how the future would pan out. My father often said, "Faith removes fear", but more than faith I would say, "Knowledge removes fear". As we know more about the future, we are more likely to be mentally prepared to negotiate and navigate our way.

Life is always in forward motion. We cannot go back to the past. As time moves on we will have to face the future. Why not embrace the future then? Why shouldn't we strive to be aware of what the future could bring? When our minds are ahead of our times, we are more likely to ride the future and maybe even shape it. Someone said that the best way to predict the future is to invent it. We all may not be able to invent every kind of change but we could all be prepared to surf and ride the waves of the future.

This book is an endeavour to look into the immediate future and the future which our little ones would navigate as adults. The rapid change over the last decade in technology is changing the world in myriad ways. The speed of change is only going to accelerate and sometimes we don't know what we need to prepare for. Many of the technologies that we may read about in this book are on the cusp of adoption while some may take more time to fructify. However, at some point in the future we are likely to come across many of these technologies being adopted and they would form an important cog in our lives.

Technologies have brought about great changes in the way we live, perceive life and interact as a culture. Whether it is television, internet, mobile or social media to name a few, all these have impacted our lives on many fronts and changed the very dimensions of our life. It has changed societies and even impacted culture. The world is a global village and technologies that are introduced in any part of the world would soon find its way into our homes. Nowadays, new technologies are quickly adopted and permeate into every corner of the globe. There is no escape. The waves of change are coming. So it is best to know it and ride it. This book is an endeavour to help you keep abreast.

The Future Beacons

"Humans are always in a state of flux, while the only constant is change, survival is by adaptation. But to thrive, one needs to actually create the future rather than just adapt to it."

– Author

"The suppression of uncomfortable ideas maybe common in religion and politics, but it is not the path to knowledge, it has no place in the endeavour of Science."

– Carl Sagan

Just like the universe itself, life in this universe actually has no boundaries. The only boundaries are the self-limiting boundaries that humans place on themselves. Once we begin to view the world on these terms, we will begin to see infinite possibilities.

The rules of present life are meant to be broken and replaced by pushing the boundaries to form new rules. These are again broken and redefined as humankind pushes the envelope and seeks new frontiers. It is but natural that the star-ship one is riding on, is grounded in science, technology and philosophies of life. As our star-ship traverses forward it seeks to always challenge the status quo. It begins to question the fundamental way things are being done at present.

This struggle of the contemporary coming to terms with the future is a constant battle in which the future will often prevail. The only thing that prevents us merging with the changed future is the human tendency to hold on to the past and present. This is done more because humans fear change and resist it. There is psychological comfort when things stay the same. This prevents humans from breaking new ground.

However, the outliers and crazies in society are likely to be the change agents and there are always going to be other early adopters who are going to embrace change early and show the way for the rest of us. This struggle has happened since early humankind and its first fuelling pit-stop was during renaissance when modern science accelerated and broke away from the shackles of religion. This battle for change will still continue.

Each chapter in this book brings out some aspects of the future which are undergoing great change and which are likely to transform our lives. The chapters are clearly titled and are written in easy to follow language. The book tries to bring insight into the waves of technology led changes which could impact us in our life time. The list may not be exhaustive but makes for interesting reading and seeks to inspire the reader to find out more about other technologies that may not be covered it this book.

This book seeks to tickle the reader's inquisitive side to know more about the future. While it would be an enjoyable read, the book also brings interesting perspectives to how change would impact businesses, the economy and our lives. Happy reading!

Chapter 1

THE PACE OF CHANGE

Sorry, change is not going to wait and it only accelerates

"Time, tide and change wait for no one. The only constant is change, embrace it or be left behind."

– Author

"Even as you think, what is imagined becomes reality."

– Author

In the year 2005, while I was doing my MBA, there was a group discussion taking place in class. The discussion was on technology and how it was changing the world. As each person predicted the next big thing, I proposed that tele-conferencing would be next. As I was from India, which was a big outsourcing destination, I recognized that tele-conferencing would be a great business opportunity. Multi-nationals often had their supply chains spread around the world. This necessitated that executives spent more time on an aeroplane going from one place to another to meet people and manage their company's businesses. As businesses at that time were going global, their executives travelled frequently to meet people who handled their supply chains, or were channel partners, customers, manufacturing partners etc. They would also travel to visit plants, manufacturing sites and other potential business partners.

In this context, to me tele-conferencing software made great sense. It would cut out unnecessary travel and save both time and money for companies.

As I spoke, in distant Luxembourg City, a company that had been incorporated in 2003 was fast gaining traction. Two visionaries *Janus Friis* from Denmark and *Niklas Zennström* from Sweden were already building such a product. The company's growth accelerated rapidly and by the year 2006 it had gained more than 100 million users who had downloaded and used the application.

The governments were so threatened by this new technology that enabled *'Voice and Video over Internet Protocol' (VVoIP)* that many of them tried to throttle, censor or regulate it. In fact, China initially banned it. Bangladesh and Oman were also against this company that had taken telecommunication to a new level. This company that had gained instant patronage by many users was *'Skype'*. 'Skype' was later acquired by *eBay* and was finally bought up by *Microsoft* in the year 2011.

Back at the classroom, I continued my discussion on telecommunication and how it would redefine the future of business and accelerate the growth of companies by facilitating a convenient platform for collaboration and enable quicker decisions that would save both money and time. While making this observation and propounding this business solution, I was completely unaware that there was this company 'Skype' that was working on the exact same problem.

I even went on to propose that since most communications were non-verbal, in the future as one engaged in discussions, one would also have 3-D holographic images of people in conference rooms. The people attending those meetings could, in reality, be seated thousands of miles away from each other in other continents. They would, however, appear real and as if seated in the meeting being held in a given location. This would make collaboration easier and no different from the case where they would actually be physically present.

In my mind, I had imagined that just like *'Princess Leia'* who appears as a holographic image to guide *'Luke'* in early *Star Wars* movies, any employee in a company could be invoked into a meeting to discuss business matters.

At that time it was just my imagination, but now this is a distinct reality. Now in the year 2019, after about a decade and a half after that imagination of mine, we actually have holographic images projected by holographic projectors. These actually blur the line between the real and the virtual. These images appear so real that people actually react to them as if it is truly part of their physical reality. In fact some 3-D Holograms are so life-like it becomes impossible for our eyes to distinguish between the real and the imagined.

On a lighter note, I wish we had this kind of technology when I was in school. I would have left my virtual recorded self to attend and take notes in class, while I spent time playing soccer in the sun. Both my teacher and I would have been happy, and my teacher would think I was taking notes diligently. I presume, these are also some of the ways technologies can spread happiness!

The Pace of Technological Change

"What is a 'Pain Point' from a Customer point of view is an 'Opportunity' from an Entrepreneur's point of view."

– Author

In the year 2016–17, I often travelled to New Delhi to meet prospective clients or investors for my start-up. In one of my visits, I stayed with my cousin in *Gurgaon*, (the place was later renamed *Gurugram*), which is a satellite township near Delhi and I caught the Metro to the city sometimes visiting Noida from thereon. The *'Delhi Metro'* was fantastic. To save money, (bootstrapping my entrepreneurial venture and saving every rupee), I often travelled by it. It was a real boon.

Though the 'Delhi Metro' was a boon especially during summer, it did come with pain points when it came to purchasing tickets. Many patrons including yours truly had to stand and wait in a long queue to purchase tickets. Even if one bought a *'Metro card'* which could be used to access the Metro, it needed to be topped up at regular intervals. This was bothersome, moreover, with every trip, travellers needed to find this card (women were rummaging into their handbags, men their wallets) to present it at every turnstile at both entry and exit. Here again, we would have a human traffic jam with people in queues trying to get in and get out. This irked me no doubt, but also got me thinking. From an entrepreneur's point of view, every problem was always an opportunity.

Most problems can be solved with some deep thinking done from ground up. All components in the environment should be looked at, as tools and solutions should be built with minimum complexity. This was my meditation and my passion. While there were many technologies to choose from, the ideal I then thought would be a method whereby the said amount is auto-debited from the user's e-accounts at entry and exit in accordance to the start and end of the customer's journey. This again seemed like the best solution that could be implemented at all Metros all over India. Why, this could be applied across any means of transport. Any person with a registered account should be able to just enter and exit ANY public transportation in a city, ANYWHERE in the country without having to bother to carry cash, purchase a ticket or use a 'Metro card' or any such system.

This could be fulfilled by reading a biometric, say an 'Iris Scan' that could be suitably placed at entry and exit points. This system would identify and charge customers to their registered accounts according to where they enter and exit the metro system. Technologies for Iris scan have long existed. Even the ability to scan a person's Iris from a greater distance had been proven possible. Hence, I began to think along those lines.

I recalled that a long time ago around June 2001, a company *'Keyhole Inc'* was incorporated and it released a very interesting software that provided the user with the ability to see high definition images of the earth from space. It went further and had the ability to create 3-D images of objects on earth. Though it was initially used for real estate, urban planning and defence, its claim to fame and commercial success was during the 2003 invasion of Iraq. It allowed *CNN* and its viewers to witness the war in a completely new way. It showed aerial footage of the war with 3-D views like never before. 'Keyhole Inc.' was then acquired by *Google* in the year 2004 which then integrated Keyhole's technology into its offering known as *'Google Earth'*. In the year 2004–05, I had to visit a place in Chicago. I had used 'Google Earth' software to simulate flying over Lake Michigan into Chicago from over the Lakeshore drive side. Flying between buildings, I finally landed on the building that I had to visit. It was fun, new and exciting.

It's important to understand what goes into the creation of this technology and its building blocks. The first challenge in creating this kind of application is obtaining or having the access to high definition images of the city from space or with aerial photography. What may be easily fulfilled by a few drones today was a lot harder in those days. In order to take a photograph from space one would need to have really sophisticated cameras. The other problem was stabilizing the camera in space which was a project in itself. However, since then, this technology has come a long way. India had launched advanced *'CartoSat'* Satellites for remote sensing which provide for multispectral high definition images of earth and observations from space. Ever since, India has been a world leader in imaging from space for civilian use. However, military satellites from the US and around the world are known for their incredible resolutions often being able to even read the fine print of a newspaper held by a reader sitting on a park bench. This technology has been prevalent for a very long time.

When I proposed that travellers' Iris be read from a long-distance I imagined along the lines of this technology. As it was already in use, it only needed to be deployed in the right context to charge a customer for using public transport. By placing suitable cameras at entry and exit points this could be easily implemented.

A little more thought and I realised an Iris scan is quite intrusive and unnecessary. Facial recognition would be a better option. This was non-intrusive and more acceptable. Hence, I started working on this idea which I thought was more feasible.

I met a distinguished gentleman in early 2018 in Bengaluru. He had co-founded *'iSpirt foundation'* and was a well-wisher who I had known for over 3 to 4 years. I thought that an entrepreneur's ideas had to be tested and bounced off a third person to get more perspective. Since, this seemed to be novel and interesting I decided to discuss the idea with him.

Well, as I ran the idea past him and began to explain how facial recognition could solve long queues in public transport he smiled. He sat back and said, "Well have you heard of *'Digi-Yatra'*?" I was taken aback and I shook my head and said that I had not. He said "the Indian Government in the coming months is going to unveil *'Digi-Yatra'* which was a plan to incorporate facial recognition at airports to process passengers in a smooth and frictionless manner". The proposal was to enable smooth passenger movement and ease the process of entry into the airport and when boarding a flight. All this would be frictionless and at any point the traveller would not need to produce a boarding card or ticket.

The Indian government, as he had said, had already drawn up plans to invite established technological start-ups to provide services to enable such a scheme. When I looked it up, the civil aviation ministry had already mooted the idea months before.

This story just illustrates the pace at which technology is being created and adopted around the world even as we think forward. When I was an MBA student more than a decade ago and I came up with thoughts about new possibilities, it often would take decades to see them turn to reality. Now, even the government seems to be forward thinking. If governments, which were considered old, boring, out-dated, slow, bureaucratic, unimaginative and useless, are adopting technologies and thinking ahead, then imagine what the private sector could be capable of doing. The pace of technology and the pace of adoption and change are going to be explosively quick compared to the past. This is going to disrupt existing orders and unleash explosive 'creative destruction' which is going to change our lives and disrupt the old ways of thinking.

Every technology transforms society and the way people live. It would even bring about cultural changes. The pace and impact of technological change is increasing at an accelerating rate. Technologies have been democratized and are easily accessible. It is now cheaper to innovate and most importantly there are a young breed of entrepreneurs who believe in possibilities. While change has always been there, the pace at which technology is catalysing the change is a new factor to contend with. It seems that truly time, tide and change wait for no one. What is imagined becomes a reality even as we think. Fasten your seat belts and brace for the ride as you read the following chapters!!

Chapter 2

HOW 'INFORMATION EXCHANGE' ACCELERATES THE PACE OF TECHNOLOGY AND CHANGE

Technology has the ability to feed on itself

"If I have seen further, it is by standing on the shoulders of Giants."

– Isaac Newton in 1675

"The pace of technological change and development is determined by the pace at which information is exchanged."

– Author

During the year 1996–97 in the midst of the internet boom, I was doing my M.S. in Engineering in the US and had gone to collect my assignment from one of my professors. As he handed it over to me, he pointed out at a sentence in my assignment. It read, **"The Pace of technological change and development is determined by the pace at which information is exchanged"**. He asked me as to whose quote it was. I said it was mine. He looked at me and then smiled. He handed me the paper and I began to ponder on the sentence which I had written on the flow.

Isaac Newton had famously said that he stood on the shoulder of giants. This is not an off-hand statement. The tremendous work done by many of his predecessors, allowed Newton to go further with his theories. Of course, none of us can take complete credit for anything we do. We have only merely progressed further from what our forerunners have left unfinished.

Over centuries, the wealth of knowledge built by several civilizations was transmitted at a much slower rate than today. Information then was often carried by traders and seafarers. They travelled great distances to exchange products often bringing back not just traded commodities but also ideas and innovations. The Arab traders not only traded with India but also carried a wealth of information and innovations back to their land and further into then Europe. Right from the concept of zero, metallurgy, astronomy, shampoo, to the refining of zinc, various other innovations had found their way from India across the globe and were carried forth by these traders.

It was never a one-way flow of information though. Cuisine from the Arab world, like *Samosas* (a baked or fried dish), the introduction of chillies and tomato from South America, *'Chinese fishing nets'* and paper from China found their way to India. The world and India benefitted greatly from the flow of information and ideas that were carried by these traders. However, it took hundreds of years for the exchange to happen and for the progress to accrue.

Fast forward to the 1990s, the internet was taking over the world by a storm. Access to information was at one's fingertips. This information superhighway was going to change the world. The exchange of information and consequent collaborations that were being built upon existing knowledge were going to benefit humanity in innumerable ways.

I remember when I was in high school we used electronic typewriters to get our projects done, and boy, it was a pain. In my final year of undergraduation in 1996, we used *Word-Perfect* to get our thesis done. While this was also painful, it was indeed a long way from typewriters that were used decades ago, when every change meant retyping the whole page again. Using *MS-Word* during my Master's thesis was such a relief. It definitely accelerated the pace at which one could complete one's work and I would often thank Bill Gates for it. Single-handedly he made it easier and quicker to do our projects, college work, and thesis. He probably saved us six months in the process!!

When the internet first came to India around 1995–96, the access was limited to a choice of text only download or an option that included images.

The government heavily controlled the internet at that time. I remember that the internet service provider was a public sector entity and that it took a really long time to connect to the internet and sometimes it didn't even happen. A gentleman from another country was sitting by me at the internet service provider's office and was suffering the same plight. We both heard the same beep, the static and some other high-frequency sounds only familiar to internet users of that generation. It truly challenged our patience. He then threw up his hands in exasperation and said that the government just wants to sit on everything. I concurred and said that they should let the private sector in, to allow for better and more efficient services. Well, that did not happen until much later.

When I think back about those days I am at awe at how patient we must have been! To connect to the internet was a prolonged affair and the connection provided was a pain compared to the instant broadband or 5G connection of today. I got access to the internet during my last year in undergrad and was trying to surf for information on universities in the US. The connection was all text, excruciatingly slow, and it often disconnected.

I remember when I first went to the United States, my mother would climb two floors to get to the computer on the top floor of our house to type out an email enquiring about how I was doing. It was the first time I had moved out of home and she was concerned. My mother would follow set pieces of instruction to start the computer, log on to the internet and then open the email service and email me. My brother-in-law would joke that I was one of the few Indian students who had an internet savvy mother.

Over the years, as technology progressed and as it was adopted, I began to see changes in India. Companies began to use the internet to hire employees, find suppliers and get customers. From travel portals, job sites to government services, many sectors and services were beginning to have an online presence. This saved a tremendous amount of time which would have otherwise been wasted in paperwork, finding, filing and snail mail. Technology truly accelerated the pace of work and improved productivity.

Productivity is a keyword. In the mid-1990s the US went through a period of accelerated growth but correspondingly lower inflation. This confounded the then chairman of the Federal Reserve. While it was difficult to deny that something profoundly different was happening it was clear that it was unlike all the business cycles since post world war. This expansion was reaching record lengths and was far stronger than expected. Growth was happening and yet inflation was subdued even in the face of tight labour markets. This completely defied logic and conventional wisdom. Alan Greenspan, Chair of US Federal Reserve, did have an explanation for it. New conceptual framework and models needed to be created to understand the new phenomenon.

A once in a lifetime acceleration of innovation had propelled the economy through the stratosphere. It was facilitated by computers that had vastly allowed for increased productivity. This brought about large leaps in growth without inflation. In the then Federal Reserve Chairman Alan Greenspan's words "The reason is that 'information innovation' lies at the root of productivity and economic growth. Its major contribution is to reduce the number of worker hours required to produce the nation's output".

To draw comparison, when my father a civil engineer by profession, needed to execute a project in the '70s, he would first have to communicate the details of the project in a typewritten paper typed in by a stenographer cum typist. This would then be checked for errors, re-typed, inserted in to envelopes with the right address, the right amount of stamps would be used and then the office assistant would then reach the post office within the stipulated time and post the same to the architect's office. The drawings were then created by hand by draughtsmen/draughtswomen who painstakingly drew every detail into chart paper on a drawing board. These drawings would then have to travel from the architect's office back to the engineer's table via snail mail to be scrutinized. Any changes would then be communicated to the architect again using the postal service, thus correspondences would continue back and forth to complete the process until the final drawing was approved. Then blueprints would then be created and the final diagrams would be sent to a government approval body for an approval permit. The government body would invariably ask for changes and this would need to be incorporated. Again these would be communicated by snail mail to the architect's office and the drawings would need to be redone by hand. This would go on and on until the architect, engineer and the government body agree on a final diagram. This was the process then, it would be long drawn and time consuming. These cumbersome processes and elongated timelines were required to just get the drawings ready even before construction could actually begin.

Fast forward to 2011, when I was working for a construction firm, the selection of the architect, specifications for the product and communications to the architect would be completed within a day via email. The drawings would be made ready by the architect within 10–14 days and communicated back to us instantly, again by email. Any changes would have a turn-around of a day or two and these would be sent as an attachment to the email. This would then be sent to a government body and submitted in CDs/flash-drives. They would check for deviations in the diagram using specialised software. Any changes and approvals thereon were done in a rapid fashion. The entire process would be compressed into a fraction of the time required otherwise. This is the power of technology, whether it is *AutoCAD*, processing software, email and now mobile devices for communication, they have all compressed the timelines for entire processes. This has given a booster shot to productivity.

Computing and communication technology have vastly accelerated the pace at which humans could make far-reaching strides and progress. In fields like biotechnology large amounts of time were spent on setting up research and data generation rather than actual analysis. Now, technology has freed up and enabled researchers in numerous ways. Bio-informatics and 'lab in a cloud' technologies have empowered researchers with tools to enable and accelerate their work in finding new solutions. These are just a few examples of technology as accelerated progress.

The **pace of exchange of information** truly accelerates the pace at which things get done and contributes to the **accelerated pace of human progress**. As communication technologies leapfrogged, so did the pace of exchange of information. Communication is now instantaneous and this has had a tremendous impact on the speed at which civilisation has progressed. As Newton said we truly stand on the shoulder of giants. **The faster we are able to exchange information, the quicker the progress**. Advances in technology and communication have enabled just that.

Chapter 3

WHAT IS YOUR BEEF ABOUT?

Why what you eat may be reclassified and how societies are going to be impacted

"Sometimes short term needs defeat long term wisdom."

– Author

"Sometimes it may be possible to reconcile short term needs with long term wisdom. The ingenuity of science and humans may make this possible where the destinations align and there is no ambiguity."

– Author

In the summer of 2017, a young 15-year old boy called Junaid Khan was stabbed to death in India by a mob which suspected him of carrying beef. The attack started when a fight broke out between Junaid, his three brothers and another group over fights related to seats in a train. They were travelling from New Delhi to Mathura and got into a fight when the other group alleged that the food packets Junaid and his brothers were carrying had beef. A group of vigilantes then set upon the brothers which resulted in the unfortunate demise of Junaid.

It is to be noted that there are some states in India that ban cow slaughter for religious or cultural reasons. While there is nothing in the law that allows for vigilantes to take the law into their own hands or dish out extrajudicial justice, some vigilantes have been practicing what can be termed as 'Beef lynching'. The more appropriate term for this would be *'Gau-athankwad'*, a portmanteau of the Hindi words for cow and terrorism.

Historical and Cultural Perspectives

In the late 90s, I was studying in the US and was invited for lunch by an Indian-American couple who were essentially *Gujaratis*. Their ancestors had migrated from Gujarat, India to Africa centuries ago. Gujarat in India is a semi-arid region with vast salt pans that placed limits on agriculture and related activities. The Gujaratis however were resourceful people who learned to do with the little they had and made the most of their advantages.

Gujarat abutted the sea and the Gujaratis over the centuries had opened another dimension to their livelihood as traders and seafarers. They traded for centuries with Arab countries and the African continent. They had strong business roots as they had to survive the limited opportunities in semi-arid Gujarat. Centuries ago, many of them migrated to Africa which was lush green and fertile. They lived there as communities and survived harsh times setting up businesses and working long hours to make it succeed.

Many of them had settled in Uganda and Kenya. By the 1950s they prospered so much on the back of their hard work and enterprise that it began to bother the locals a lot. Anti-Indian sentiments erupted here and there. By 1960s this unrest grew and was bolstered with political overtones in these countries. It was then that the many Indians who had settled in Africa started to use their 'British Protected Person Status' to look at the UK as an alternative. In Kenya, Africanization happened over a course of time. In Uganda, things were more fluid. At the height of the xenophobia, Asians including Gujaratis were given 90 days to leave Uganda. Many Gujaratis who had settled there thought of leaving to India. India, on the other hand, had a refugee crisis of its own with the refugees from Bangladesh flooding India during the war for the liberation of Bangladesh. India was focused on the war with Pakistan and was trying to liberate Bangladesh. Pre-occupied with the needs of its internal security and that of its immediate neighbour, India could not offer help. Moreover, it was unclear as to which country these people were citizens of and who should take them in.

Dictator Idi Amin of Uganda expelled the Indians and Asians alike, as he believed they did not invest back in society. However, within a few years of their leaving it became apparent that the immigrants were not sapping the system but had been driving growth and prosperity. Shop fronts were uncared for and vast agricultural lands turned to jungle as the enterprise of the expelled Asians had left with them. Inflation soared and growth tanked. Three decades later, the government did invite the Asians back but only a smaller percentage came back.

Meanwhile, many Gujaratis had used their British Protected Person Status and had headed to the UK to finally settle in Leicester and London. Over the following years, they re-established businesses from scratch and began to flourish all over again. After having spent some time in the U.K, some went on further west and migrated to the US.

I was at a dinner hosted by one such couple. They had settled in the US and had brought up their two daughters to be American citizens. The daughters were attending college and had adopted the American lifestyle. As we spoke over dinner the gentleman decided to test my response to the current topics of discussion that were being debated in India. He tried the hot button topic of Kashmir first and it fell through. I agreed with him that it is a subject that would be best put into the cold freeze for 100 years while India pursued development. The future generations would have greater wisdom and perspective to resolve the issue as they would be less emotionally affected by issues that muddle it, I added.

The next hot button topic he picked was cow slaughter. In response, I said that cows were considered wealth for centuries in India and a person was often described as so and so landlord who had these many acres of land and this number of cattle, etc. He said it was the same in Africa too. In fact, people would describe their wealth in cattle and often eligible bachelors would receive cows as dowry. In response, I said that I was aware of that and mentioned that I knew a Kenyan-American in the US who had mixed feelings about this tradition. While he did not want to dismiss the ancestral traditions his family insisted on, he could not agree with them. He had mentioned that his fiancée was modern and born in America. While he did not want to disrespect his parents, he was not sure where he would house the cattle he would eventually be given as dowry!!

Jokes apart, having established that cattle were considered wealth, I began to explain to the Gujarati gentleman how the humble cow became holy in the 'Cradle of Civilization'.

The Cradle of Civilization

While vast swathes of land in India are fertile, India has always been heavily dependent on the monsoon rainfalls to recharge its reservoirs of water that sustain agriculture and life-stock. These may include lakes, ponds, check dams, the underground water table and even feeders to non-perennial rivers. For thousands of years, the monsoons have maintained a pattern of rainfall that repeats itself with varying degrees of fulfilment. The Indian civilization has flourished for centuries nurtured by these abundant water sources. While humans are said to have first appeared in Africa, India is often termed as the 'Cradle of Civilization'. Behind the success of a long uninterrupted inning is the availability of this critical element which we call water. It is vital and is most necessary to sustain life.

While the monsoons did seem to work like clockwork, they would at times fail or sometimes come up deficient. There have been prolonged phases in history where the deficient monsoons have tested the very survival of this great civilization.

How Cows Became 'Holy'

Cow's milk and products derived from it are integral to Indian cooking. Whether it was milk, ghee, *paneer (cottage cheese)* or yoghurt, Indians used these bounties to their fullest extent. The problem, however, was that vast swathes of land in India were monsoon dependent. While the monsoons were normally regular and periodical, they sometimes failed or were insufficient. There were prolonged periods of drought that would not only affect agriculture and its output, but also affect livestock. Cattle were as integral to survival as water and it only made sense to not deplete cattle numbers when drought-afflicted the land.

When there was a severe drought, the best of cattle would be the first to be sacrificed to sustain livelihood as they fetched the best price. In such circumstances it was difficult to provide for them and they would be more valuable as meat. However, the problem was that it was the 'best' of the livestock that was often sold off, as they fetched the best prices. This caused enormous problems going forward and would be poor livestock management. When the drought conditions regressed, the cattle population struggled to recover, moreover the best of the cattle breed had been sold. This would deteriorate the quality of the livestock. To prevent the 'killing of the golden goose' something had to be done at a societal level.

It was however impossible for the learned and informed folks to reason with farmers when survival during the drought was on the top of their minds. No matter the depth of their wisdom, the short term needs tend to defeat long term wisdom. This was the genesis of how the cow became 'Holy'.

Embedding the 'Holy Cow' in the Cultural Ethos and Heritage

To really put this in perspective I will narrate a story that recently occurred. This is a true story about how the Indian coastline along the state of Gujarat became the birthing place of the large whale shark. This story is contemporary and would give the reader a glimpse into how practicality and rationality have to be woven into cultural ethos to bring about positive results.

The slaughter of whale sharks along the Indian coast went on for over half a century. Initially, it was not against the law to catch them. However, today the whale sharks have many well-wishers among the fishing community in the state of Gujarat. How did this happen? The whale shark traverses across the oceans and in summer it arrives off the coast of Gujarat to give birth to pups. Until the 1980s the whale shark was hunted for its many uses. The oil from its liver was used to waterproof the hulls of the wooden boats. The Gujaratis have been traditionally sea-faring and these were important applications that could be derived from the whale shark. Moreover, in the 1990s East Asia paid top dollars for shark fins and these became prized exports among the fishermen.

Exporters also traded meat, skin, oil, and cartilage of the whale shark. This only accelerated the killings of the whale shark taking it to the brink of extinction. With so much money to be made only GOD could save these gentle giants.

Mechanized boats were used to hunt and kill these magnificent sea dwellers. Hundreds were killed and their killings only accelerated even as their populations dwindled. A short film 'Shores of Silence' by Mike Pandey highlighted the issues related to the killing of these whale sharks. The Wildlife Trust of India (WTI) subsequently lobbied hard and the Indian government brought the species under the Indian wildlife protection act in 2001. It was the first fish with such a distinction.

However, even years after the enactment of the law, the fishermen had barely heard of it. Also they were not bothered by a distant law nor did they care about the remote possibilities of prosecution. Short term needs overrode the consequences. It was then that a programme was started and launched by Morari Bapu, a popular spiritual leader and bard. In a publicized event, Bapu went out to sea at Dwarka and blessed a tangled whale and set it free. He said that the whale shark was like a pregnant daughter returning home to give birth. His metaphor caught the imagination of the common folks and the fisherman promised not to hunt them.

From then on even if the fishermen accidentally netted the whale shark, they cut their nets to let the shark loose again. The Gujarat Forest Department played its part and compensated the fishermen for every damaged net. Over the years, hundreds of whale sharks were liberated and the department disbursed hundreds of thousands of dollars as compensation for damaged nets all in an effort to protect the whale shark.

The compensation did not cover the whole costs of a damaged net. Neither did it compensate the fisherman for time lost and time taken to set it free. Yet, the coastal communities who were instrumental in the killings earlier, had now turned into saviours. Once part of the problem, they became part of the solution. As its profile was raised the creature was adopted by seven cities in Gujarat as a mascot. In fact, there is a 'Gujarat Whale Shark Day' celebrated on the New Moon Day of the ninth month of the lunar calendar.

This story goes on to illustrate how positive change in society can be brought about by first understanding an issue and giving it the correct type of treatment. The whale shark's story is similar to how the cow became 'Holy'. According to the unique circumstances of this vast sub-continent the survivability of the Indian civilization hinged on proper livestock management. Livestock management was necessary for the survival of the best of breeds and cattle through the drought. Considering how the cow was integral to the life of agricultural India, it was embedded to be 'Holy' in the cultural ethos and thus over centuries the objective of protecting the species was achieved. Bestowing holy protection upon cattle and weaving it into the cultural sensibilities of the Indian population ensured the survival of India as a civilization. As funny or out of place it may seem to many western observers there was a sound basis for the cow being 'Holy' in the Indian context.

On a lighter vein, my grandmother would talk of her own mother rearing cows in the backyard of her home in rural India. They had named the cud chewing bovine 'Lakshmi'. She said it provided milk for the entire family and extended family. My grandmother had told me it was an integral part of their family. She said that by treating the animal well, by calling it by a given name and showering it with love, the cow actually gave more milk. When my grandmother mentioned this, I did not believe this story then. Years later, I read that cows that were given names and addressed by the same, had higher yields of milk than the cows that were not given love and were treated badly. Well, it surely is a 'Holy' cow!

Vegetarianism and the Trade-Offs

My wife and I would debate on vegetarianism and I would tell her that there are large numbers of domesticated animals in the world only because many humans have found uses for these animals dead or alive. This is not to sound very stone hearted but to just state an observation. I did not mean to shoot down my wife's strong beliefs in vegetarianism but to highlight that humans would only rear animals if there was some benefit they could derive from rearing them.

Chickens, for instance, may have been as rare as peacocks in the wild if humans did not see value in raising them for their meat or for their eggs. So there are trade-offs in this debate on vegetarianism or veganism. I just thought I will slide this argument in for people to understand that economics and sustainability often dictate the course of survival of species as much as compassion for them might.

Clean Meat

In the meanwhile, advances in science have turned the whole concept of 'farm to fork' on its head. Especially, when it comes to animal meat in the form of in-vitro animal products, the whole system is about to undergo a sea of change. In-vitro animal products are produced from stem cells harvested via biopsy from living stock. These are then grown in the labs over weeks to produce what is termed as 'Clean meat'.

This kind of process will eliminate cruelty to animals. In fact, the animal rights organization 'PETA' has been investing in in-vitro meat research for years in the hopes of a commercial breakthrough. It has even offered a huge reward of 1 million dollars to the lab or scientist that would be the first to take in-vitro chicken meat to the market.

Reduce Snow Storms in Your Area?

"Clean meat is clean in more ways than you can imagine."

– Author

The mass adoption of 'clean meat' could have many unexpected benefits and positive offshoots. Interestingly, the change in consumer habits could actually reduce snowstorms in your area. You may be justified in asking "What is the connection?" Actually, there is a connection. 'Clean meat' production is much cleaner not just from a 'cruelty to animals standpoint; it is also cleaner from an environmental standpoint.

'Clean meat' can be produced without the use of antibiotics, without leaving a large environmental footprint, without excess contamination or the cruelty associated with animal slaughter. The meat industry produces more greenhouse emissions than the entire transport sector. For instance, it is estimated that 27 Kilos of CO_2 equivalent of greenhouse emissions are emitted to produce 1 kilo of beef. To put it in terms of something we can relate to, it is equal to the emissions equivalent of driving 63 car miles. That's massive positive savings as we move towards 'clean meat'.

Apart from the fact that 'clean meat' will ultimately eliminate cruelty and killing of animals for food, there are other enormous environmental benefits. 'Clean meat' is environmentally sound and humanely produced and can also be called 'guiltless meat'. Many environmentalists believe that this method to produce meat could give a booster shot to the fight against global warming.

While challenges in regulatory approval, costs of production, communication issues and challenges in changing consumer habits remain, the technology and science behind 'clean meat' is unstoppable. Over time, it would become the preferred way people consume meat. It could play a key role in reducing harmful greenhouse emission significantly. This technology could be a key weapon in our fight against global warming and even ensure survivability of the human race.

There are very many positive offshoots from this science. Apart from the prospect of saving many potential victims like Junaid, this technology could well save the environment and thus humanity!!

Chapter 4

THE FUTURE OF EDUCATION

Why technology and communication would bring about a
'Tectonic shift' in the way education is delivered

"Imagination is greater than Knowledge."

– Albert Einstein

"For an education system to be a success, it needs to address 3 major factors; relevance, cost, and delivery."

– Author

"Technology has the ability to democratize education, to spark explosive growth and bring about progress by tapping into the potential of every human mind across the world."

– Author

The Dawn of Education

The earliest written texts were written on birch bark or palm leaves and were later written on parchment. Among them, The *Vedas* from India are considered the oldest of the texts dating back to over two millenniums B.C. These books of knowledge were vast compilations which reached their peak in periods spanning the mid-2nd and mid-1st millennium B.C. during the iron and bronze ages.

The transmissions of the Vedic texts were oral by tradition with the precision and accuracy maintained by elaborate mnemonic and chanting techniques. These texts were mostly in *Sanskrit*, widely considered the mother of languages, along with *Tamil* which is considered the oldest living language in the world. Bearing in mind how vast this body of knowledge is and how consistently it had been transmitted over millenniums, it is widely believed that texts were transmitted in both oral and written form.

The problem however with the oral tradition and limitations of handwritten texts were that it could not be widely dissipated. Since knowledge was transmitted orally one needed to attend a *'Gurukula'* or school run by a 'Guru' to partake in the dissemination of knowledge. This meant that knowledge did not spread easily and was confined to a smaller privileged group. This led to the *'caste system'* in India, where knowledge was passed on from one generation to the other. This then became an inheritance or birth based system. While there were attempts to reach a larger audience with the establishment of universities in that era, it did not have the mass effect desired. The oldest of these universities were established at *Nalanda*, *Vikramshila*, and *Taxila* in as early as the sixth century B.C. However, these did not help in breaking the monopoly over knowledge held by a few privileged people.

While the original objective of the caste system in India was to encourage specializations and prevent a SINGLE group from enjoying all privileges of society namely 'Power', 'Knowledge' and 'Wealth', it led to major problems. While one group was to enjoy 'Power', the second group was to enjoy 'Knowledge' and the third group was to enjoy 'Wealth', no assigned group was to enjoy all the three and have a monopoly over society. While on the face of it, it seemed to be a system where the societal advantages were NOT held by only ONE group of people, the system also led to the formation of groups that neither enjoyed 'Power', 'Knowledge' nor 'Wealth' and were bereft as victims of oppression. Moreover, it failed to allow a person to choose his profession based on his/her natural proclivity and talents. This innately was a huge disservice to society and in the long run, it did not allow for the society to harness the strengths of its inhabitants to the fullest extent.

For instance, the knowledge of metallurgy was confined to a closed group of blacksmiths and transmitted orally to the long line of blacksmiths in the family. There were three key problems that this created. The first is that the knowledge was easily lost over the centuries; the second was that without the ability to disseminate the knowledge, it was not built on or furthered with the collective knowledge or wisdom of the larger population; the third was that, if the knowledge was to be kept confined within a group, mass production of goods was not possible. All these factors led to the demise of an amazing golden period in the history of the world where several inventions and innovations were created and many of them perished due to the limitations brought on by the social hierarchies.

While there are many evils in the caste system this aspect alone is highlighted to show how society limited progress and hurt itself. When the ability to transmit knowledge to the masses is hindered, it precludes the abilities of the larger populace and its capacity to evolve and develop a larger knowledge base for the advancement of human kind. Education by nature has to be disseminated to a larger group who could gain and later contribute to it. This is why the 'Printing Press' was a ground breaking innovation that was as important as the invention of the wheel or the discovery of fire. It broke a major limitation that hindered the sharing of knowledge and laid the foundation for the further progress of humankind. This was a small step by man but a giant leap for humankind.

The Revolution Called the 'Printing-Press'

One of the most powerful and revolutionary inventions in the second millennium was the printing press. The invention of the printing press allowed for knowledge to be widely disseminated and laid the base for great strides to be made in the progress of humanity.

The early technologies to print were prevalent since the *Tang dynasty*. During the early part of the millennium movable type printing had been invented during the *Song Dynasty* in China and later used during the *Goryeo Dynasty* in Korea where metal movable type printing was used as early as the 13th Century. Woodblock printing based on screw presses were widely used in Europe during the 14th Century. Until then the metal movable type printing technology had been unknown in Europe.

A goldsmith by the name Johannes Gutenberg adapted existing technologies, innovated and perfected his innovation to mechanize the printing press. It was developed around the year 1440 A.D. and this revolutionized the spread of knowledge and information. Within several decades of the invention, it spread to hundreds of cities in Europe. Its reach went further and the output from printing presses soon touched hundreds of millions of copies.

As printing presses became popular, sufficient paper was required to be made available. Paper manufacturing processes needed improvement and new innovations mechanized the manufacturing methods. The introduction of water-powered paper mills replaced handcrafted techniques used by Chinese papermakers. The mechanization of paper making and improvements in printing technology stoked the 'print revolution'.

The Foundation for the 'Education Revolution'

It won't be far-fetched to say that the *Renaissance* in Europe had been fuelled by the arrival of the printing press. A new medium of expression and mass communication called "The Press" had come into being. This altered the very structure and fabric of society in Europe. It threatened the established political and religious order and set the stage for mass reformation.

Literacy spread and broke the stranglehold of the literate elite who until then had domineered over the masses. Further, the 19th Century ushered in the era of the steam-powered rotary press, which allowed for mass industrial scale printing, thus fuelling the spread of knowledge and information. This accelerated the progress made in science, technology and literature which caused the 'education revolution' over the next two centuries.

A number of 'Land Grant' Universities were set up in the United States and the state began to fund education in the United Kingdom in the 19th Century. While early education became more democratized, college education was still elusive. There were still some barriers. Books were printed but were expensive; libraries had memberships that were exclusive. The knowledge that was available was in books, but its dissemination, the cost of training and the cost of education were still barriers.

'The Information Super Highway'

I would often say that the coming of the internet was a new dawn. The Information superhighway which was built on the back of the internet boom allowed for information to be dissipated over the world-wide-web across the globe to reach people in far off regions who were otherwise disconnected. Most of the world was now armed with the tools for connectivity and had access to information which until then they had been excluded from. Well YES, in terms of access to this knowledge, the world truly began to shrink.

Vast bodies of information were now available to anyone who had a web connection to search and find. The establishment of search engines such as *Google* truly allowed for information on the web to be indexed so that they could be searched for and found. It had truly revolutionized what was impossible.

The possibilities are many and over time they have only increased manifold having gained infinite dimensions. Nevertheless, the mind-sets toward how education for all can be achieved would take time to fructify. For instance, in India, even when information today is available at a click of a mouse, education for the first part of the decade of the 21st Century had still been anchored to rote learning. This may be part of the remnants of the past oral traditions. Subjects which were of great significance in the 19th and 20th Century such as physics, chemistry, and elements of mathematics were still given heavy weightage. Though these were important pillars that supported the scientific revolution, now in the 21st Century, there is no reason to have a syllabus anchored by them alone.

It is to be noted that the world has moved into another era. We are well into the internet and mobile revolution. Any information on any subject can be got in a jiffy and there is no value in storing all the data and information in one's head. Rather than memorizing without comprehension, efforts can be better channelled towards training people to find information, collate, analyse and draw deductions from the informational data. Moreover, subjects relevant in the past two centuries need to be upgraded to weigh in new subjects of relevance in this century including Computers, Programming, Environmental Sciences, Personal Finance, Investments, Ethics, Social and Civic responsibility, Humanities, Art and Culture, Artificial intelligence, Analytics, Robotics and so on. We need to move further and offer choices beyond the traditional subjects. Children should be given a broader choice beyond the usual and be able to pick their interests.

The 'Mechanics' of Absorbing Information

An interesting aspect of the way learning was done in the past and what's relevant today is its mechanics. For instance, my mathematics teacher who taught the 11th and 12th grade would teach a concept and then tell us to practice hundreds and thousands of problems in that concept. Very often, the only purpose gained by doing the hundreds and thousands of problems was enhanced pattern recognition abilities. The set mind-set that accrued from this did very little to further creativity or innovative thinking. This system only served as training for an exam which was born out of a system that was outdated and torturous.

During my study in the Unites States of America, the first thing I noticed was the difference in what was stressed. Unlike in India where the pressure was on mastering what was already there, the assignments in the United States focused on research and furthering of one's knowledge and creativity. This actually made sense even from an economic point of view. There is very little value in trying to master what has already been mastered. Any corporation or start-up would value what is new and effective rather than what they already know. In business, we call that 'differentiation'. The ability to produce products that are different from competitors is valuable, marketable and it would certainly command a premium versus something that is already 'commoditized'. It is interesting to note that the success of the United States of America over the last 100 years has been because of this kind of innovative thinking which focuses on thinking differently. This truly has added value. While in India the stress was on mastering and perfecting what was already there, the U.S. saw value in furthering the knowledge by invention and creativity. This was clearly reflected in their education systems and consequent economic progress.

In this new modern internet age the world is overflowing with abundant information and knowledge that is easily accessible, hence there is greater value in the ability to find the right kind of information, research and to further build on its worth. The internet era has truly transformed the way information needs to be handled and worked on. It is this 'mechanics of learning' that will take time to seep into the system so that society can make full use of it. Today innovation and creativity have a premium over knowledge and it will remain this way for the conceivable future. The world has truly changed.

Democratization of Education

The future truly lies in the 'Democratisation' of information. Low costs and ease of delivery of knowledge form the pillars of democratization. The online media through text, video, and audio provides a powerful outreach making delivery possible to the remotest corners. In India, a large number of people are beginning to see the smartphone revolution take off. It is said that when the per capita income of a country crosses $3000 (2019 numbers) and the number of people using the smartphone crosses 500 million a powerful force will be awakened. These numbers are coming to reality as far as India is concerned and the internet revolution powered by the smartphone is beginning to take hold. Many people in India are experiencing the internet for the first time through their smartphone, they are leapfrogging ahead as they skip the customary computer terminals. This has the ability to empower and democratize education and impart several skills in a way never seen before.

Online Courses

While these things may be true of India the world is undergoing a quiet revolution of its own. Higher education in the form of online degrees is beginning to see a future. This method is both low cost and has outreach. People can learn from their homes, at a time of their own convenience and gain the necessary skills to be employable in the market place. For those who find university along with boarding and lodging unaffordable, new avenues are being created to achieve learning and for building skill.

The key advantage of creating educational software/online educational tools is that the only major cost incurred is the cost of creating the course content. Delivering and replicating it even a million times for millions of applicants would only cost cents to the dollar. It does not matter how many people it would be delivered to, the costs would become incrementally negligible. When spread over a large number of people, the costs are greatly reduced and affordable. Technology enables the democratization of education by keeping costs low and ensuring instant delivery that is far-reaching.

'Micro-Credits'

'Micro-credits' are going to be the future. Aspiring students would be able to take any chosen course online, free of cost at a time and place of their choosing. They would only need to pay for exams to test their learning and obtain the certificate proving that they have taken that particular course. In this manner, learners can accumulate 'micro' credits which can finally lead to a degree at some time in the future.

The advantage of such a system is that a student could choose courses he/she may find relevant and it could be across subjects. In this way one can have a valuable and unique bag of skills across the spectrum making one more employable rather than being confined to a set syllabus. Cross-functional knowledge and skills are often highly valued in new age companies. A multi-disciplinary approach can offer skills tailor made for modern requirements and make one more employable. One may for instance, choose to do a course in spread-sheets and programming while doing another in equities and along the way one could gain micro-credits in both. This would make one instantly employable with those institutions or companies looking for those specific skills. One could then do further courses on the side as one works on a job and accumulates credits. If for instance, one wants to change one's job one could do a course in statistics on one hand, marketing studies and data analytics on the other. This would sufficiently skill one to become a market analyst. In this way, one could be equipped to take on a different stream, expanding the opportunities and options available to one. This way one could keep accumulating skills and be job ready for the market and continue to stay relevant to a changing job market. In one word this system serves the third pillar in a successful education system which is RELEVANCE.

By having relevant course work which is accessible through online courses and that is delivered instantaneously and remotely at a fraction of the cost, we would create a system which truly democratises education and empowers the masses.

A number of micro-credits when accumulated could lead to a degree over time. This system is not only LOW COST but also convenient with easy modes of delivery and replication enabled by new technologies. The courses can also be short and RELEVANT to the job, saving both time and money for the aspiring applicant. Companies would list requirements in terms of skill sets and micro-credits rather than degrees, giving them in turn, the specific skill sets they want rather than a broad degree. This would be win-win for the applicant and the companies.

Learning in the future would be over a lifetime in tune with changing requirements of the industry. One could start earning early as one accumulates relevant skills and moves up the value chain in one's given broad horizon of interest. This would be done across established fields without being bottled down into a particularly narrow focus. This would give one the ability to build unique skills across various fields of study giving every individual a unique perspective from which one could contribute effectively. Moreover, there would be better job-skill match from an employer's point of view. Education would be affordable and relevant to the requirements on hand while allowing students to earn as they learn and move their skill sets upward in a flexible fashion. This would also be a flexible response to the requirements of the market place.

The future of education is going to be shaped step by step in the form of micro-credits and engaging real world work experience. Most education is going to be delivered online at affordable costs in self-paced learning modules. Low cost, multi-modal delivery and relevance are going to be its defining pillars. People are going to pick skills one micro-credit at a time even as their study is interspaced with work, it would culminate into a degree over time.

After all, small drops do make the ocean!

Chapter 5

ELECTRIC VEHICLES

How EVs (Electric Vehicles) are going to transform the very assumptions we have made about transportation

"It is not that people fail to see change, they only fail to believe it."

– *Author*

"Paradigm shifts happen and the automobile industry is going to go through its biggest shift yet by 2025."

– *Author*

"Gasoline mixed with air, broken up into a mist and partially vaporized is ignited in the combustion chamber and this causes the mixture to ignite and the gases to heat-up and expand as the I.C. engine goes through its 'power stroke'. In the 'power stroke', the piston moves downward and in the process rotates the crankshaft through the connecting rod. Please feel free to interrupt with any questions you may have", the professor said as he continued explaining about the internal combustion engine or the I.C. engine as it is widely known. However, in my mind I was thinking that we are busy mastering what is already there, what if there was a better way to power a vehicle?

'Paradigm Shift'

Since, my early years of engineering under-graduation, there haven't been any phenomenal changes in the automobile field. However, change is imminent. During my Masters in the US, my 'Major Professor', who was a quality control expert, played a video in which the father of quality control Edward Deming was speaking. He was talking about how a *'Paradigm shift'* could change the landscape. He gave the example of the carburettor manufacturers and suppliers who were producing better and better carburettors, yet the carburettor manufacturers went out of business. This was because there came along a better technology; the fuel injector. While the carburettor manufacturers were perfecting the carburettor they ultimately went out of business because they failed to pay heed to the winds of change. There was no point producing a better and better carburettor when a 'Paradigm Shift' was imminent. When the fuel injector came along the carburettor manufacturers were caught on the wrong foot and went out of business. Now the question is "Are we nearing a 'Paradigm Shift' again and in a much bigger form?"

The 'Kodak Moment'

In the financial world, we have something they call the *'Minsky Moment'*. It is a moment of realization as a sudden collapse occurs after an extended bullish run. I would draw a parallel in the free market economies for product manufacturers. I call it the *'Kodak Moment'*. It is the moment (which needs to be framed!) when the CEO of the major global manufacturing company realized that the product they had invested in and perfected over the years suddenly had no market as a new wave of technology swept his company aside. You can embrace it and surf it, or fight it and drown. The choice is yours. The wave when compelling is unstoppable.

That moment of realization is the 'Kodak Moment' in deference to what happened to the Kodak film company a few decades back. As I was watching the Edward Deming video described above, the Kodak film company was going through an existential crisis, the root cause of which was similar to the 'problem in the thought process' described by Deming in the video. It all happened around the same time. The shift had begun. Now at this moment in time as I write this book another shift of tectonic proportions is going to shake the automotive world and oil industry. It is even going to change the balance of power as oil-producing countries will have to stop depending on oil revenues and do real work, while oil-consuming countries will attain energy security and free up resources to be used in other productive sectors of the economy.

The whole oil trade economy is going to be shaken up and new players could take center-stage in the automotive world. It's a new dawn. Embrace it or be left behind.

The 'Tipping Point'

Many people may wonder when it would be that moment in time when EVs (Electric Vehicle) takes the baton from the ICE (Internal Combustion Engines). Just as when the sun emerges from behind the thick clouds and there is sudden bright light, the much awaited transition would be unexpectedly soon. I would go on to call such a moment the 'Kodak Moment' of the automotive industry. Just as the film cameras disappeared from the market with the dawn of the digital camera we are going to see the disappearance of the ICE with a sudden wave of EV in force.

My first interaction with an EV was when I drove the EV1, which was General Motors early attempt at electric vehicles. It was quite an experience when I drove it around the Milford Proving grounds in the late 1990s. This was the time I was at my first paid job at General Motors in the United States. EVs weren't the 'in' thing then but now they are coming into their own. However, for the EV to take over the market completely from the ICE some essential factors need to fall in place.

There are some key metrics that need to be reached for the wave of technologies to be compelling and for mass adoption to come into effect. When those metrics are hit we would see a tipping point and the consumer would have no reason to buy any other product other than an EV.

Key Metrics

Some of the 'Key Metrics' one needs to consider are the following

- Upfront Cost of a Vehicle
- The Range of the Vehicle
- Battery Life of the Power Source
- Speed of Charging

- Maintenance Costs
- Running and Fuel Costs
- Resale Value
- Support from respective Governments

Let us analyse these factors one by one.

Upfront Cost of a Vehicle

While the cost of Lithium-ion batteries was $1,000 per kWh in 2010, the costs had whittled down to 1/5th by 2017 and are still falling now. It is expected that the cost of the Lithium-ion batteries would fall to $100 per kWh by 2020. This is a key metric and a key tipping point as at $100 per kWh, the upfront cost of an EV begins to become comparable with an ICE vehicle (Internal Combustion Engine vehicle). The price would continue to drop from there on thus making an EV a compelling proposition and the vehicle of choice.

The Range of the Vehicle

The range of the Lithium-ion batteries is directly related to its energy density. The energy density of the batteries has been increasing at a rate of 5–8% per annum. It is estimated that the range of a small to a mid-size vehicle could hit 500 km–1000 km per charge. In fact, the *Tesla Roadster* which launches in 2020 has a stated range of 1000 km. If that is the benchmark range for an EV, it is the ICE vehicles that will begin to be dwarfed and their range would be subpar. The EVs are on course to pass the test in flying colours.

Battery Life of the Power Source

It has been estimated that the batteries lose only about 1% of their capacity for every 30,000 km run. This is a stunning metric in favour of EVs. A brand new ICE vehicle drops in fuel efficiency at a much faster rate than an Electric Vehicle. Everyone knows that with mechanical wear and tear the ICE engine begins to drop fuel efficiency at a much faster rate. This in itself makes the electric vehicle a preferred option. This also means that the upfront costs depreciate much slower and the lower 'Total Cost of Ownership' of an EV is compelling.

Speed of Charging

Solid state batteries using sulphide super-ionic conductors are beginning to show promise. These batteries can operate at super capacitor levels and completely charge or discharge in minutes. It is by far more stable and safer than current batteries. Similarly, Graphene batteries are being researched in right earnest and hold great promise. While there are still challenges, the promise held by such technologies will make supercharging a reality soon.

Maintenance Costs

A typical ICE vehicle can have anywhere between 2000–4500 moving parts while an EV can have as little as 20–160 moving parts depending on various other factors. The contrast is obvious. With fewer moving parts there is lesser wear and tear and therefore fewer parts to replace. Interestingly, the cost of repairing an EV has been found to be far cheaper than the corresponding ICE engine. Hence the lifetime cycle cost of an EV as compared with an ICE Vehicle would be significantly less.

Moreover, the EV can keep running for eternity and it would not need any emission tests, engine oil change, nor has it to be compulsorily scrapped after a certain age as in many countries. All these show that the EV trumps the ICE vehicle in lower maintenance costs and is a clear winner.

Running and Fuel Costs

The EV derives its power from a charge in its battery. Hence, the only cost of powering an EV is the cost of electricity. It has been estimated that with the constantly fluctuating price of gasoline/diesel which powers most vehicles and the comparable cost of electricity, the cost of running an EV would be as little as $1/7^{th}$ the cost of an ICE vehicle. These metrics would revolutionize transportation and bring down the cost of many commodities whose major cost component is just fuel and transportation. It would actually significantly bring down inflation in many countries. It would revolutionize trade and shrink the world further.

Resale Value

The resale value of an ICE vehicle is likely to collapse very soon, as EVs come into the foray and start taking hold of the market, people would especially stop buying used or second hand ICE vehicles. Moreover, the EVs are not going to depreciate as fast as an ICE vehicle, as they do not wear out fast and will hold up in the resale market. This phenomenon is going to accelerate the demise of ICE vehicles and is going to collapse the demand for such vehicles. When all the metrics become compelling and the ducks line up, we are going to see the sudden demise of the ICE vehicles in a matter of a few years. It is really going to be another 'Kodak Moment'.

Support from Respective Governments

The biggest potential car markets in the world are the US, China, and India. China has already announced incentives for EVs and has mandated by law that any vehicle manufacturer has to obtain an electric vehicle score of 20% by 2025 up from 10% in 2019. This has given a great push from the government side. In 2018, the Prime Minister of India, Mr. Narendra Modi has set a vision to bring about changes to transform India into an all-electric car market by the year 2030. The Indian government has even changed laws to even enable individuals to sell electricity through charging points in the cities. This was done by changing regulations that previously permitted only state utilities to sell power. India has also formulated policies to offer subsidies for the purchase of EV vehicles. The Indian government has also proposed to only allow electric two wheelers to be sold in the two wheeler market after 2025. The Indian government has also dropped the GST (Goods and Services Tax) on Electric Vehicles from 12% to 5% in order to incentivise the sector. In the US the federal IRS offers tax credits for purchasers of EVs. There is a credit of $2500-$7500 per EV according to the range and battery capacity for the first 200,000 vehicles sold per manufacturer.

Conclusion

EVs are an unstoppable force and most people are going to be quite surprised when the swift shift actually happens. It is going to be a 'Kodak Moment' that needs to be filmed for posterity. In the coming years, once the key metrics for the EVs are achieved, it is going to be a tsunami and we are going to see a churn in the market place and a tectonic shift in world power dynamics. Those who believe in it will capture it and ride the advantage, the rest will drown.

Years ago, when we bought our Honda Sedan my uncle who had worked for years in oil industry came to visit. He looked at the car and commented that it would probably be the last of the gasoline-powered vehicles we were going to buy. I did not believe him then but I do believe him NOW!

Chapter 6

AUTOMATED FLYING TRANSPORT AND DRIVERLESS CARS

How automated transport is going to bring about a revolution in the transportation industry

"To be chauffeured around is a privilege today, but it will be the norm tomorrow."

– Author

"The only permissions needed for change is feasibility and economics."

– Author

Even in the early part of the last century, horse-drawn carriages were ubiquitous. Then automobiles began to take over. India, however, has always been unique. India can best be described as a large dinosaur with its head in the future and a long tail into the past. Modes of transport which were relics in most parts of the world, often stayed on in India much after they were phased out in the rest of the world. My mother spoke of a horse-drawn 'Tonga', locally called a 'Jutka' that stood at a particular junction, many decades ago. While I was at kindergarten, I vaguely remember seeing a 'Tonga' ply students to school. This mode of transport had survived time in India and plied in the midst of cars and other modern transport to ultimately disappear, altogether. In the United States, the Amish people can be seen in their horse-drawn carriages even today.

I have brought this up to make a point that, just as the horse-drawn carriages were ubiquitous and seemed part of the landscape more than a century ago, cars as we know them today, seem inseparable from the landscape of today. However, just as the cars of today completely replaced the carriages of yesteryears we are likely to see sweeping changes in the near future where the cars of today would be completely replaced by vehicles of the future. Just as the carriages of yesteryears were reduced to novelties, so would the driver-driven cars of today. In the future, autonomous cars and autonomous air taxis would completely replace the driver-driven cars of today. The driver-driven cars would be reduced to recreational vehicles. Unimaginable you think? Well, read on.

Autonomous Cars

People would have seen the commercial advertisements where Volkswagen Polo brakes automatically and where the Ford Focus automatically parallel parks. These use proximity sensors which are becoming increasingly common parking aids. If you take these technologies and combine it with technologies such as automatic steering technology and technologies such as cruise control, we begin to see the shaping up of the very building blocks of an autonomous vehicle.

It is well known that *Waymo LLC* a subsidiary of *Alphabet Inc*, the parent company of Google has been giving a push for driverless technology. A typical Google autonomous test vehicle has over 8 sensors including the very noticeable roof-top *LIDAR*. The LIDAR is a camera that uses an array of 32 to 64 lasers to measure the distance of objects, to make up a 3D map over a range of 200m. These give the vehicle a picture of the obstacles ahead and also indicate of any potential hazards. A standard camera also points through the windshield and acts as another set of eyes for the car. Bumper mounted radars keep track of vehicles in front and behind. The car also has a rear mounted aerial that receives geo-location information from GPS satellites. There are also ultrasonic sensors at the wheels that monitor the car's movements.

Together, these sensors and cameras look for any nearby hazards like pedestrians, cyclists, stray animals, other motorists, traffic ahead and around.

The sensors can even read street signs and detect traffic signals. Data that is received and collated by Google's central software from its array of sensors and radars in the vehicle are combined to identify existing or potential obstacles. These are also used to read common road signs and to manoeuvre accordingly. The data also helps to direct the vehicle towards its destination, keeping it safely on the road and away from obstacles. The software can also understand potential actions that could be taken by other motorists by observing and capturing signs such as flashing side direction indicators or even a cyclist extending his hand to turn. The Google software then directs the autonomous vehicle to slow down and gives way even as it re-adjusts its own speed and direction in anticipation of actions by other occupants of the road.

Internally, the car is equipped with a tachometer, gyroscopes, altimeter, and accelerometer to give positions of the car's movements. These instruments add to the highly accurate data that is captured relative to the movements of the vehicle. These help the vehicle to operate safely and within the set boundaries or rules of what would be considered safe driving.

Impact of Driverless Cars

It is intended that apart from being just a convenience, 'driverless cars' would also help in ensuring fewer accidents and safer driving conditions. In the US alone, thousands of people are killed in car accidents each year, additionally many more are injured because of human error. Worldwide, the number of deaths caused by road accidents is actually in the millions. This is a rather sobering picture for road safety. These statistics buttress the argument for removal human error from transportation and speak for a future with autonomous vehicles.

Driverless cars will not only lead to lesser accidents, they would also revolutionize car ownership. Along with the trend towards a 'shared economy', people would just hail an autonomous vehicle rather than own a car. It is not as though people would give up car ownership. It is just that car ownership would come down drastically.

Once upon a time owning a hunting rifle was a necessity for hunters to hunt for food. Nowadays, people own hunting rifles for recreation rather than to hunt for food. Similarly, there is going to come a time when people would own cars for recreational driving rather than day to day needs which would be easily achieved by hailing an autonomous taxi. Such are the changes in attitudes that these new and innovative technologies are going to bring towards car ownership.

Interestingly with fewer accidents and fewer claims, the insurance companies that insure cars and drivers are going to be impacted. Along with the advent of EVs (Electric Vehicles) and the fact that less number of accidents are likely to occur, the business of automobile spares and replacement parts would shrink. These changes may drastically do away with some of the verticals in many of the automobile businesses. These developments in technology would affect the future of vehicle insurance and the business of vehicle spares which would slowly become less and less lucrative.

Automated Flying Transport

Traditionally in the game of cricket, a batsman would be advised to keep the ball along the ground as he played his stroke. This was to avoid getting caught out and to prevent the loss of a wicket. However, with the advent of the shorter format of the game, scoring quickly became the Holy Grail and many batsmen began to take the aerial route, clearing the fielders and often the boundaries. Adam Gilchrist, Australia's legendary wicket-keeper batsman, Virender Sehwag, the Indian dynamo, and Sanath Jayasuriya, the Sri Lankan legend, famously believed "there was more room in the air" than on the ground.

Staying true to that maxim it is only natural that entrepreneurs and innovators take to the sky to relieve congestion on the road. While flying cars were the stuff of science fiction for years and have been dreamed about in even mythology, they are coming to your neighbourhood not long from now.

Mythology

Early Indian mythology talks about *'Vimana'* or 'Flying chariots' described in Sanskrit Epics. The Sanskrit word 'Vi-mana' means "measuring out or traversing". 'Vimana' stood for self-moving aerial cars in mythology. Most famous among them is the *'Pushpaka Vimana'* of *Lord Kubera. King Ravana* from the *Ramayana* took the 'Pushpaka Vimana' from Lord Kubera. He is said to have used it to kidnap *Sita* and to take her to Lanka. Thus began the great battle to retrieve Sita which eventually led to *Rama* defeating Ravana. Rama is then supposed to have returned the 'Pushpaka Vimana' back to Kubera.

Interestingly, when parallels are drawn with the 'Puspaka Vimana', it was nothing but an autonomous flying vehicle which flew to any destination along with its occupants on command. An interesting feature of the Vimana was that "it always had a seat that was empty". The moment someone came and occupied the seat another empty seat would appear. So, the vehicle technically had unlimited capacity. Now, how cool was that!

In Our Times

In our times, considering the congestion on road, it becomes obvious that it may be a better idea to look skywards rather than dig tunnels (Sorry! Elon Musk). This is where autonomous flying vehicles seems to be an interesting bet. However, like any other new introduction, regulatory and operational hurdles remain. Nevertheless, these issues would be sorted out with time.

Work in this direction has already started and a German start-up is piloting the world's first Autonomous Air Taxi (AAT) in Dubai and has signed a contract for that with the government of Dubai. This Volocopter formerly known as the e-volo can cruise at 50 km/hr. and has a flying time of 30 min. This is a rather good start. As commuters take to the skies, it would be interesting to take note the Volocopter is powered by batteries, which put to rest the concerns about excessive pollution. As ATT's are powered by clean sources of energy and their batteries are charged with solar power, they would be a boon and would not only solve the problems of congestion and long waits in traffic, but also provide for a more environmentally friendly mode of transport that does not pollute or contribute to global warming.

Conclusion

Automated flying transport and driverless cars are the future. We are going to see more of these in the near future. While driverless cars are going to make the roads safer, automated flying transport is going to help relieve urban congestion and even redefine our urban spaces.

Well, as they say in cricket there is more room in the air. Keeping with the shifting paradigms, the future would be one where we look up to the skies!!

Chapter 7

TRANSPORT FOR HIRE

How people are going away from 'owning' to 'renting' transport and what this entails for society and the automobile industry

"Nobody in this world actually owns anything. Philosophically speaking we are just temporary dwellers and are renting everything including our time on this planet."

– Author

"Though the natural tendency for humans is to possess and own stuff, the younger generation prefers to rent. The economics and convenience are just too compelling."

– Author

In the year 2011, a young Indian company was trying to raise capital and two angel investors from India, Rehan Yar Khan and Anupam Mittal were evaluating it. The company was an online marketplace for cabs and rental services. When Rehan approached other angels to co-invest, many didn't see value in the business. They said that in the city they resided in, they could just wave their hand and hail a cab. Why would they want to search online for it? Hence, Rehan ended up writing most of the cheque for the Angel round investment.

Fast forward to 2015, when subsequent rounds of funding were raised Rehan's 100K investment was worth about $36 million dollars. At that point, his stake was 1.6%. The company was now valued at over $2 billion dollars. The company in question was *Ola* which became the rival to *Uber* in India.

The Economics of Transport for Hire

Years ago, while I was in B-School, we looked into an interesting case study about *South-West Airline* in the United States of America. The success story of South-West Airline in an industry that often courted losses became a case study in a number of B-Schools. There today is a corresponding airline called *Indigo* in India which can also be considered as a no-frills or frugal carrier. The secret sauce to their success was simple, "Keep the aircraft flying most of the time and limit time on the ground".

Indigo passengers would notice that everything about the airline is geared towards turning over the aircraft and getting it ready to take-off for its next outbound destination. The aircraft does not accommodate late passengers. They board and de-board the passengers in the quickest time possible. The aircraft clean-up begins while the aircraft is still in air, before landing commences. The air hostesses are also trained to collect the trash and dispose the same. The baggage handlers and clean-up crew are ready to go as soon as the plane arrives. The aircraft does minimum catering. They take off and land on the dot and everything is organized to finish like timed pit stops of racing cars at a grand prix.

This meticulous attention to detail actually has an economic motive that may not be immediately obvious but it surely makes a world of a difference. A business that gains returns from the running of its equipment would get greater return with greater utilization of its machinery. In the case of the aircraft industry, the major capital invested is towards its aircraft. The longer it stays in air, taking more and more passengers from point A to point B, the greater would be its revenues. When the equipment's non-productive hours and downtime is greater, not only is the business not making revenue, but the equipment is also depreciating with time without contributing to the top-line. When the equipment is expensive, having significant downtime can lead to a significant hole in the balance sheet and income statement. Profitability is tied to maximizing the utilization of that equipment.

Now substitute the aircraft for cab services. The same principles apply for cab aggregation services. Rather than scouting for hours looking for rides, cab services could run continuously, picking up passenger B, as soon as Passenger A disembarks. When this is done without long gaps, it ensures longer uptime and better utilization of the cab itself, which would drive up profitability. Further, the proximity factor has greatly helped many cab drivers, where the cab's next ride is close to the drop point of the previous ride. In this manner, cabs save not only time but also fuel thus maximizing the load factor. The use of technology, where the potential customers of a cab are matched to the nearest cab, increases convenience for both the customer and the cab driver. Factors like pinpointing pick up and drop off points on a map and ability to make cashless payments makes the entire system a sure-fire winner.

How often have you seen the cabs of the past, either just sitting around or driving around the block looking for passengers? Though it is true that in places like New York and Mumbai a cab could be hailed by just waving your hand, this is not true for many other cities and suburbs. The business model for an app based cab aggregator service is a clear winner with the advantages it brings to the table.

Changing Behaviour of Customers

Anand Mahindra the CEO of *'Mahindra & Mahindra'* a major automobile manufacturer in India was talking about the future of the automobile business and he talked about 'shared mobility' and how the Ubers and Olas of the world are changing the very dynamics of the automotive business. He was clear in his mind that the automobile sales around the world are going to stagnate or shrink because of the factors that we had just discussed. While he felt people are still going to buy cars for intercity transport and recreational purposes, he felt that Ubers of the world are going to revolutionize transport within cities.

Also, it is to be noted that in general, if the utilization factor of the vehicles on the road go up with the shared economy, there would be lesser need for vehicles. Interestingly, Anand Mahindra had also said that his daughters preferred hiring to owning vehicles and did not even bother to learn how to drive. This has been the outlook of the new generation. They would rather rent than own.

One of the disadvantages of ownership of machinery is extended unutilised idle time. When one owns a car, at most, one is likely to drive it 2–4 hours a day.

A cab service would be able to keep the car running for a minimum of 16 hours a day in 2 shifts. The utilization factors become apparent. This allows them to operate at fares that are irresistible. A rider, who hires such a shared service on a need basis, need not put down capital or pay instalments, buy insurance, go to the fuel bunk to tank up regularly, pay parking charges, clean and maintain the car, make oil changes or attend to repairs. S/He could just tap his/her phone and be chauffeured around in a 'pay as you go' model. Plus payments could also be done in a cashless fashion. Who wouldn't take up such an offer?

Future of Transport

While some like to stay grounded, some like to fly high. Improvement in battery technologies has brought up the potential for not only road transport but also the potential use of drone-based taxi services. With autonomous vehicles slowly gaining technological momentum, drones that are pilotless are the next in line. Already cities like Dubai are experimenting with a flying drone taxi service that could be hailed using a mobile app.

A German start-up is piloting the world's first *Autonomous Air Taxi* (AAT) in Dubai and has signed a contract for that with the government in Dubai. This *Volocopter* formerly known as the *e-volo* can cruise at 50 km/hr. and have a flying time of 30 min. It is only 2 meters in height with a diameter of the propeller rim extending to 7 meters. It has nine independent battery systems and comes along with an emergency parachute. New companies such as *'Uber'* and *'Quadcopters'* are interested in the *VTOL technologies* (Vertical Take-off and Landing), and the future is ripe with possibilities. However, these services must be rolled out with caution as the technology will face legislative and operational challenges, which need to be ironed out before it can come into mainstream use. Yet this is an interesting glimpse into the future.

Conclusion

The future is not going to be like anything we have seen before. It is going to be straight out of a science fiction movie. Flying taxis and autonomous vehicles would take us where we want to go at the tap of a mobile app, or who knows, maybe even on just a thought!

Chapter 8
SOLAR AND CLEAN ENERGY

How and why solar power and other clean energy sources are going to transform the future of energy

"Sun is the source of all life and energy and any life form that grows and moves on earth has its origins of energy from the Sun."

– Author

"Perennial availability, cost effectiveness, environmental friendliness, and easy accessibility will propel solar power as the preferred choice of power generation for generations to come."

– Author

For many centuries various cultures have worshipped the 'Sun God' as the source of all life and energy. In their ancient wisdom the Indians knew that the Sun's energy was the backbone of all life on the planet and there would be no life without the sun. It is not just the east Indians but also Egyptians and Meso-Americans who had developed their religion around the Sun God.

Many rulers claimed to be incarnations or descendants of the Sun God. The ruler of Peru was one such example. Even in Japan the Sun Goddess *'Amaterasu'* played an important role in ancient mythology and was an important deity of the imperial clan. To this day the sun symbol represents the Japanese state. After all, Japan is the 'land of the rising sun'.

Even in ancient Roman history sun worship was prominent. The feast of the unconquered sun called the *'Sol Invictus'* on December 25 was celebrated with great joy. This date was eventually celebrated as Christmas day, and as the day that Christ was born.

Symbols of the sun were common and a recurring theme in Indo-Iranian, Greco-Roman and even Scandinavian mythology. While in Egypt the Sun God was worshipped as *'Ra'*, the Indians of North America celebrated the sun with the famous *'Sun Dance'*.

In India, the Sun God was worshipped as *'Surya Devata'* (*Aditya, Bhaskara* etc.) and glorified in the Vedas. He (anthropomorphised as male), was considered as the source of all energy and life, which drove away disease and dispelled darkness. Many Indian kings and clans held that they were direct descendants from the sun. This is not actually some mumbo jumbo. In fact, all life on earth is because of the sun. In some sense, we in reality are all descendants of the sun.

Source of Life on Earth

As taught in elementary classes plants need the sun for *'photosynthesis'*. Plants have a green pigment called *'chlorophyll'* which resides in the 'chloroplasts' of plants. 'chlorophyll' is a key ingredient which enables 'photosynthesis'.

During 'photosynthesis' the plants takes in carbon dioxide from the atmosphere and water from the ground. It then captures the sun's energy and converts all this into sugary carbohydrates and releases oxygen in the process. Plants are the greatest consumers of solar energy, trapping the energy like a charging battery and converting it into plant matter. Hence the sun's energy is captured by plants as it grows, entrapping it into various forms of glucose and organic matter. We can consider plants as batteries which store the sun's energy on earth. Plants are called producers as they produce food for animals to eat.

Herbivorous animals eat these plants and convert the plant energy into protein, fat, glycogen and energy. When consumed, plants are digested, absorbed and converted into animal protein, glycogen and fat. Some parts of it are used by the animal in their day to day activities as carbohydrates. As the animal uses their store of fat and glycogen they release carbon dioxide and energy. This is the reverse of what the plant did to grow as a source of food. In the presence of sunlight, plants absorb carbon dioxide and create organic matter from the carbon while releasing oxygen in the process. Humans on the other hand, release the energy obtained from consuming organic plant matter and expel carbon as carbon dioxide, after breathing in oxygen. In this way animals can be called consumers.

When *Carnivores* or meat eaters consume animals they absorb animal protein and fat. They assimilate it and convert these into fat and protein of their own. Finally, they release the energy stored in these and expel the carbon in the form of carbon dioxide. Again they are consumers. The only difference being that they are further up the 'food chain'.

Algae and fungi then decompose dying matter and release the minerals necessary for plants to grow. This entire food chain can interestingly be seen as a transfer of energy from the sun to the plants and then to the herbivores and then to the carnivores. The plants capture a fraction of the energy of the sun while the herbivores capture a smaller fraction from plants. A smaller fraction from the herbivores accrues to the carnivorous animals. We need more energy from the sun to keep this cycle going. It becomes obvious that for the food chain to be balanced there needs to be more plants than herbivores and more herbivores than carnivores, as any food pyramid would show. Interestingly, it is to be noted that the energy we use to accomplish our day to day activities is ultimately solar energy. The food chain is nothing but a transfer of energy from the sun finally to us. Hence in some ways, we could say that we are all descendants of the sun or Sun God!

It will be interesting to note that even petroleum and gasoline based products that drive most of the world's transport are actually just storehouses of the sun's energy. These fossil fuels including coal and petroleum were created by nothing but dead plant and animal matter that were buried. The buried matter came under high pressure and temperature over centuries and converted to petroleum and gas, which was trapped under the surface of the Earth. Over the last century, these have been explored, found, extracted and exploited. Since the industrial age these sources of energy have laid the foundation of the modern industrialised societies we see today.

Solar Power and Clean Energy Forms

All forms of energy ultimately have their source mostly from the sun. Wind energy is one such form and is nothing but energy caused by the movement of air that moves due to the pressure differential caused by the sun's heat. Most of this is because of the differential heating of land and water on Earth which changes air pressure and thus causes the movement of air. Hence, wind is again solar energy in another form. An increase in man-made emissions increases the atmospheric entrapment of the heat energy from the sun. When this converts to wind energy it unleashes the fiercest storms that we have come to experience in recent times. The storms, typhoons, cyclones have been unusually fierce and this can be attributed to increased greenhouse gases that lock in the heat from the sun. This entrapment of heat energy by the atmosphere is a phenomenon we have come to call 'global warming'.

Solar energy in itself is the most direct method in which we can obtain energy from the sun. The technology to tap solar power has been around for long but has caught the fancy of many researchers now as it has become imperative to find a source of renewable power that can power the planet without causing 'global warming'.

Initially, in its first avatar solar power was used to power calculators and to heat water. It has now moved from solar power cookers and rooftop installations to be used at an industrial scale. A rapid shift from experimental generation of solar power to large commercial solar power plants is taking place. However, in the past, the greatest problem had always been financial viability. When compared with alternatives, specifically thermal power plants, the total cost of generating solar power had remained more than the cost of producing power using thermal sources. Another problem with solar energy is that it could only be produced while the sun was out and the production of solar energy dropped off during night. We then had to tackle issues such as storage of power and uninterrupted supply at night. To a great extent these problems have been solved but there needs to be more progress made in battery and other associated technologies. Moreover, solar panels occupy large land areas which may not always be available at suitable places to trap the sun's energy.

The 'Tipping Point'

"The sun is going to bring in a new dawn which is going to power us for centuries to come."

— Author

Hence, there are three major metrics that have to be satisfied for solar power to become main-stream. Firstly, photovoltaic cells on the panels need to be made more efficient wherein a greater amount of power is captured in a smaller area. Otherwise, we may need large tracts of barren land to install several panels; alternatively floating panels on water bodies are also being explored. Interestingly in India, in certain states, solar panels were placed as a cover over water channels and served a dual purpose. They prevented the evaporation of water while the solar panels produced electricity.

Secondly, we still needed to improve battery technology to store energy, for nightfall and rainy days. Here investments in technologies, supply chains and production facilities particularly for Lithium ion (Li-ion) batteries have to be improved. Now scalable solution for storage and release of electricity exists. Further newer promising battery technologies are being developed. One that holds great promise is a water-based battery that is being developed at Stanford University. This could compensate the grid when the demand peaks at night and also store energy in the batteries during the day when solar energy production is at its peak. All these technologies are improving by leaps and bounds as we speak and it is only a matter of time before we see a *Tipping Point* in its mainstream adoption.

Finally, adoption of these technologies would boil down to economics. Thus we need to look at the cost of production of solar power. The *Tipping Point* has been reached for this metric also where the cost of producing solar power as of now has become cheaper than the cost of producing power using thermal sources. This is especially true when we consider the total lifecycle cost of a thermal power plant. In fact, reliable sources point to the fact that the cost of producing one megawatt-hour of solar power in North America is only $50 as on 2018 and has fallen from highs of $350 in 2010. This can be seen against the backdrop of the cost of production of power from traditional sources such as coal which is around $100 per megawatt-hour. (Calculations are based on the

Levelized Cost of Energy (LCOE) Analysis). It is even more impressive that the solar panels today are so efficient that they don't even need 'beach weather' to operate and can operate efficiently even in Northern European countries and Canada. In fact, they become less efficient at higher temperatures when they overheat and solar energy gets converted to heat more than electricity.

Future of Solar Power

An impressive project that has developed in India in the field of solar energy is the use of solar power to power water pumps in farms. These have a number of advantages including the fact that more water gets pumped on hotter days when water is most needed. The most interesting thing however is that when the pumps are not in use, the solar sources that are connected to the grid can be used to supply the grid with electricity in a process called 'reverse metering'. Hence, excess power from the panels can be sold to the power distribution companies. This earns extra income for the farmers and also helps them pay off the infrastructure costs for the solar panels.

Another interesting concept that has come to the fore is V2G or Vehicle to Grid. It is estimated that private vehicles are parked and not in use 80% of the time. In that time EV or electric vehicles can be hooked on for charging. While there is nothing new about this, if we consider the batteries of the EVs to be store houses of energy, then the power grid to which they are connected can be managed in such a way that it can draw power from the batteries of the vehicles during peak demand periods and help to charge them during lean periods. In other words, when the EV vehicle is at office and plugged in, it charges during the day drawing power from the grid during the time the solar panels are at their peak generation capacity. However, when the vehicle is back at home at night and plugged in, it serves as a store house of power and charges the grid back in reverse during the time the grid generation capacity is limited. In this way the V2G concept can be used for power grid management and the batteries can serve as store houses of excess energy.

In the future energy for space missions could come from tapping solar energy and solar power satellites which would use phase locked microwave beams or laser emitting beams to transmit solar energy to colonies or other locations in space. In 2008, scientists were able to send a 20 watt microwave signal to an island in Hawaii from a mountain in Maui. Since then, Mitsubishi and others have formed a team in order to place satellites in orbit which could generate upto 1 gigawatt of energy. This $21 billion project is the first step to create space assets that could capture and wirelessly transmit energy from these space based assets.

Conclusion

We are fast converging to the 'Tipping Point' in alternative energy especially solar power and it is the end of the road for new thermal power plants around the world. This is driven by solid economics and not just because of environmental considerations. The sun is going to bring in a new dawn which is going to power us for centuries to come!

Chapter 9

ON-DEMAND MANUFACTURING AND 3D PRINTING

How advances in manufacturing are going to be able to convert thoughts/concepts into product that are made to order and how it would herald the future of on-demand manufacturing

"The journey from the intangible thought to the tangible product is going to happen in a jiffy."

– Author

"Maybe an aggressive prediction but it is just matter of time before we just take a bio-print of a person, almost akin to cloning."

– Author

Traditional Ways of the Publishing Industry

Traditionally in the publishing industry, an author would be given a signing amount and contracted to write a book by a publishing house or an author would write a book and approach a number of publishing houses, sometimes via a publishing agent. After the book is written and edited, the interior design for the book would be worked on. Once the cover design and all other aspects have been taken care, the book would go to print. In the printing press the printing plate is prepared and a set number of copies of the book are printed often numbering into the thousands. The books are then sent to book stores where they are stocked and displayed. The books may sit there for weeks and months and are often returned or sold at a discount to move the inventory.

Now this business model has been around for decades but changes in the model had to happen. As usual it was technology that brought about the changes. In the old model, investments had to be made by a publishing house to print and distribute the books, locking in capital in inventory, which sat on the book shelves for months on end. The 'cash cycle' of the model was very long and this impacted the business. Moreover, they were no guarantees that the books were sold. On the other hand, if the demand for the books exceeded the print, then another set number of books were sent for printing. It was tricky to match the demand and supply of the books in question. This model had many flaws.

How 'On-Demand Printing' and e-Commerce Revolutionised the Printing Industry

The new e-commerce based model uses the power of 'on-demand digital printing', online listing, social media, logistics providers and e-payments to accomplish the task of reaching the book to the reader. While advertising of the book is done through social media, a preview of the book is given through e-commerce websites. Ordering of the book can be done from the comfort of one's home and payments can be made through digital means. The book is then delivered right at one's doorstep.

The most important cog in the entire eco-system as far as the new model is concerned is 'on-demand printing'. 'On-Demand printing' along with digital technologies applied to the publishing field allow for an order of even a single book to be printed and delivered. One could order one book or a thousand books. The books are printed only on demand and not pre-printed. This is efficient and prevents the build-up of inventory and the locking-in of capital. From the author's standpoint it empowers him/her to create a book without need to either invest upfront or enter a contract with a publishing house that would have to make heavy investments upfront. If you think this concept is an isolated case think again. This entire model is being replicated across industries.

"Zero-Inventory" Model

Jewellery has been popular among women for centuries. It makes them feel good and worthy. Men have gifted women jewellery to impress them, formalise relationships or mark anniversaries. No marriage or engagement is complete without a gold, diamond or platinum ring. Traditionally, one would go to a jewellery store and try on jewellery or peruse through the entire collection before making that purchase. It goes without saying that this is changing. Niche e-commerce companies in the jewellery segment are making their mark.

There is an area in Chennai, India, known for jewellery and *silk sarees* called *T.Nagar*. No wedding shopping in the entire state is complete without a visit to this place. In fact, in one survey, the retail space in T.Nagar was rated as one of the world's highest grossers of revenue per square feet of retail space. This does not surprise me as we Indians are known to splurge on weddings and this is the place to get prepared and to get all decked up. While traditional methods of buying jewellery are still the mainstay, new changes in customer behaviour are beginning to take root.

There are select niche e-commerce players in the jewellery segment in India including on *Caratlane*, *Bluestone* and *Voylla*. It would be interesting for many consumers to note that some of these companies actually have no or very little inventory of jewellery and what may be displayed on the website are only images. In fact, the process of making the selected jewellery may only start after the customer places an order. In other words, it is a "Zero-Inventory" model. This is really the future of not just jewellery but also manufacturing in general. All this is made possible by 3D printing and advanced additive manufacturing techniques.

In this kind of jewellery catalogue, in reality, many designs are only displayed online but none have actually been made before order, stored as inventory, shelved or physically displayed. Sometimes, the site may allow for the customer to make his/her own designs or customisations. When the customer places his/her order through the online portal, the production order is sent to the production floor. Here 3D printing techniques are used to make the jewellery. Gold is literally printed from gold dust and a binder is used to form the jewellery as it is printed layer by layer. This is *additive manufacturing* technology at its best. Finally, precious stones may be placed to complete the piece before it is couriered to the customer. This is not just some *JIT* (Just in time) model this is a *"Zero-Inventory"* model.

Such technologies in 3D printing are not just for jewellery they extend to many manufactured goods. It is not long before one could send one's car for repair and the service agent orders the required spare part and 3D prints it. The agent would only get the design of the part from the car company and he would go on to 3D print it at the service station. This would not only save transit time but also money, it would do away with the need to stock several types of spares for several car models.

Many products can now be made using this 3D printing technology. Proven technology exists to make, 3D printed machine parts, handguns (alarmingly!) and even homes. 3D print technology was even used to print a bridge part by part.

The Many Dimensions of the 3D Technology

There are many out of the earth applications for this technology. Rockets going into space now need not carry bulky tool kits or emergency spares. For example, Astronauts at the Space Station could just download the design of a required tool or part, necessary to make a repair from the base station and 3D print it. This would eliminate the need to carry spares or bulky repair tools into space. 3D printing has evolved to the extent that the technology is even applicable to organic matter.

Printing of food using 3D printers has been something of interest and we have companies developing 3D printed burgers and even pizzas. Imagine how much fun it would be for an Astronaut in a space station to have pizzas made and delivered to him in space via 3D printing. It would truly be an 'out of the world' experience.

It wouldn't be long before we had 3D food printers at home just like one has a microwave or oven! It would make, well, "print" us the choicest of foods on command. The era of cooking would be past.

3D Printing also extends to medical applications and is called *'Bio-printing'*. This upcoming technology can be used to print organs such as kidney, liver, bladder and heart for patients in need of a transplant. A 3D liver was recently successfully printed and work is on as you read to make bio-printed organ transplant a reality.

3D printing has also made strides in dentistry where it is used to create prosthesis, tools and even dental implants. New methods are being developed to create many devices like 3D printed braces used in teeth alignment, 3D printed implants, dentures, dental crowns, aligners and many orthodontic appliances. With some oral scanning and CAD design, one can print anything three dimensional as long as its dimensions have been captured. Dentistry dedicated 3D scanners and 3D printers have been developed and the field of dentistry has been one of the very early adopters of the 3D printing technology for good reasons.

3D printing technology and such additive manufacturing techniques are best suited to fields such as dentistry and medicine where high levels of customizations are required. Any prosthesis or implant for example has to match with the patient's morphology which is unique. It is possible with 3D scanning and 3D printing technology to capture accurately the 3D model of the patient's dentition for better treatment of the underlying problem at hand. Apart from being accurate and customised, the technology is cost-effective and time saving. This technology is in the process of completely replacing the earlier process of using injection moulding, which has been the mainstay of dentistry for decades.

Conclusion

3D printing and advanced manufacturing is going to democratise manufacturing and make the global supply chains redundant. Massive savings in logistics and fuel are also going to make such technologies environmentally friendly. Also advances in bio-printing and food printing have a potential to revolutionise medicine and the fast food industry respectively.

Possibilities are immense. One may for instance, be able to make and eat all the burgers one wants, according to one's specification on-demand, without having to visit a burger joint! Further, if one develops cardiovascular diseases from eating all those burgers, one may just have the doctor bio-print a new heart and fix the plumbing. The future sure looks decadent and delicious!!

Chapter 10

THE FUTURE OF COMMERCE, FROM E-COMMERCE TO K-COMMERCE

When whatever you need will be manufactured in your neighbourhood and delivered to you within hours.

"The more things change the more they remain the same. Life always seems to come a full circle."

– Author

"From Mom-Pop stores to Organised Retail, from Organised Retail to e-Commerce, from e-commerce back to Mom-Pop stores, commerce in the world is going to come a full circle."

– Author

Mahatma Gandhi believed in self-sufficiency and was often found in a meditative state spinning his 'Charkha' (A 'Spinning wheel' meant to convert raw cotton into thread). He basically spun his own cloth. In those days, India was under British control. The British systematically destroyed the Indian weaving and textile industry and even went to the extent of chopping the thumbs of Indian weavers in some parts of the country as a deterrent. The British encouraged farmers to produce cotton and dyes which were then transported to England and places like Manchester where mills would make cloth which were sold back to the Indians at exorbitant rates. British then wanted India to only be a source for raw material and a market for finished products. Mahatma Gandhi launched the civil disobedience movement to counter this exploitation. It was a movement meant to awaken India and the *'Charkha'* or *'Spinning wheel'* became a symbol of self-reliance and defiance.

Mahatma Gandhi launched the *'Satyagraha'* which in Sanskrit means "holding firmly to the truth". The term 'Satyagraha' was a coined and developed by Mahatma Gandhi and deployed in the Indian independence movement. It also deeply influenced Martin Luther King Jr. in his civil rights campaigns in the United States of America. Mahatma Gandhi called 'Satyagraha' 'love-force' or 'soul-force'. The bedrock of Satyagraha was non-violence or *'Ahimsa'*. 'Ahimsa' is a Sanskrit word for 'non-violence'. The idea was to wean one's opponent from error by patience and compassion. 'Ahimsa' was the means and the 'Truth' was the end.

Civil disobedience and non-cooperation that was practised under 'Satyagraha' was meant to change the British behaviour in a non-violent way and prevent exploitation of the subjects under their watch.

Mahatma Gandhi had then visualised that India would be a country with self-contained villages that produced everything it needed. Its needs would be produced and consumed in harmony with nature. Mahatma Gandhi visualised India as a land with a number of self-contained units. The 'Charkha' was that symbol of self-reliance.

Coming a Full Circle

The idea of 'self-contained villages' seemed an utopian concept to say the least and India and the world have panned out far differently from the vision of the Mahatma. Mahatma was instrumental in the British ceding control of India back to the Indians. However, at least initially, he seemed to have called it wrong in his vision for India as a nation of many 'self-contained villages'. The world itself is now evolved and grown into a global village! Nowadays, manufactured goods criss-cross countries and continents and are made available by smooth functioning supply chains that span these continents.

Commerce is a huge behemoth in action where goods are shipped not only trans-Atlantic but also trans-Pacific. However, the future it seems could actually be something different altogether. The very concept of 'self-contained villages' could come back to roost as technological breakthroughs in *'Advanced manufacturing'* and *'3-D printing'* take hold. Could Mahatma Gandhi have been actually right in his vision? Are things coming back a full circle? Read on!

The 'Kamadhenu Effect'

'Kamadhenu' also known as *'Surabhi'* is a mythological divine bovine-goddess described in Hinduism, as the mother of all cows. She is supposed to be a miraculous "cow of plenty" which can provide its owner with whatever he/she desires. Interestingly, there was a super market chain which was called 'Kamadhenu' in my neighbourhood.

The 'Kamadhenu effect' though is when one's 'neighbourhood store' acts as the all providing 'Kamadhenu' and caters to all of one's needs. This store would not be just an ordinary store. It would be a "printing store" with a very advanced 3D printing machine running on advanced additive manufacturing technologies. It would be run by small entrepreneurs or franchise owners.

So, one would order goods on one of the online portals or apps by selecting one's choices from the store's online catalogue. There would also be portals where one could customise one's designs or even build from scratch (with assistance), according to one's needs. Depending on one's delivery location the design specifics of the selected good and the final design would be sent to the neighbourhood 3D printing store and voila, it would be 3D printed to one's specifications. Even customisation of infinite variations can be catered to. Finally the goods within hours of the order would be delivered to the home of the customer without a fuss or additional shipping charges. Add drone technology to the delivery mechanism and you have a complete the eco-system.

Well this seems to be a dream, but the technology and systems are slowly taking shape and will make this a reality. It is not very far in the future when Mahatma Gandhi's vision of the 'self-contained villages' actually begin to play out and become a reality. Well yes, life can come to a full circle thanks to technology. Welcome to the world of *'K-Commerce'* or *'Kamadhenu-Commerce'*.

Mom-Pop Stores: The Comeback Kid

Walmart is a behemoth and the world's largest company in revenue. While it is headquartered in Bentoville, Arkansas its operations span 27 countries under 55 different names. At the time this book was being written It operated over 11,227 stores and generated revenues close to half a trillion dollars every year. Though it was incorporated as Walmart only in October 31, 1969, Sam Walton had founded the company in 1962. It is the world's largest company by revenue and number one company in the Fortune 500 list of companies by revenue.

Even as the juggernaut Walmart expanded into being the world's largest corporation many critics worried about its effect on small businesses and local communities. In one such study Kenneth Stone, a professor at Iowa State University found that small towns suffered immensely. Within 10 years of Walmart establishing in a town, the small retailers often lost over half their retail trade market share to Walmart. Walmart did have an impact on small towns and local businesses and even tax collections.

However, not everything is negative. Walmart is great for consumers as it supplies all goods under one roof in a hypermarket format at rock bottom prices. I remember when I first came to the United States to study my Master's in Engineering, my American roommate and I were debating the best kind of stores to shop at. My roommate was particularly unhappy with Walmart. He said that it was driving small 'Mom and Pop' stores out of business through the strength of its size and its ability to squeeze suppliers, employees and competitors alike. He also went on to say that they were unstoppable and had already driven a number of Mom-Pop stores out of business. I, on the other hand, was just dazzled as to how big and how inexpensive its stores were.

Again this is not a judgement on big retailers like Walmart. They have been wonderful for consumers, often weaving an intricate supply chain across the globe and stocking over a lakh product items (10 lakh = 1 million) per store, bringing to the customers decent quality goods at prices that were unheard of. They have been a boon for manufacturers who were able to scale up volumes and not have to deal with intermediaries and rent seeking distributors.

While these 'Multi-Brand Hypermarket' stores have eaten large market share in the past, in the future the tides are likely to shift and the markets would change and we are likely to witness a churn again. The small entrepreneur owning mom and pop stores and smaller e-commerce companies are set to make a comeback. They would be empowered by advanced technologies and new models of doing business. With business economics on their side, they are poised to ride the shift in market dynamics. Note, I have said entrepreneurs and not just businessmen/businesswoman. These people would be more than just businessmen/businesswomen. They would be enterprising youth who see opportunity and seek independence and self-sufficiency. They would use technology in ways unheard of and tailor products and services according to the local market. They would do away with mass production that was based on "one size fits all" model. They would also stop mass manufacturing in its tracks and make complex supply chains of large multi-national corporations redundant.

Niche e-commerce sites would be born that would be more like design houses. These would be unlike the present day market place e-commerce companies. These would only provide designs, parameters or specifications including material and form factors. They would allow for customisations and even allow customers to build their own designs using tools on their site. On making an order, the selected designs would be sent to one's nearest neighbourhood mom-pop store electronically. These stores would be proximate to one's address and would be manned by enterprising entrepreneurs. These entrepreneurs would convert the intangible designs into tangible products using 3D printing and advanced additive manufacturing techniques in quick time. The parts would then be assembled to form the final product and it would possibly be delivered via drones or last mile delivery personnel. Any entrepreneur with a design sense and understanding of the process and its limitations would be able to set up a niche e-commerce design portal and direct his/her designs to a slew of neighbourhood 3D printing stores depending on the location of the customers. E-commerce and manufacturing would be truly democratised and it would not require heavy investment. Many small entrepreneurs would become empowered to deliver standard quality products at reasonable prices with prompt delivery that would be unmatched. Add to this the simplified logistics and environmental benefits of lowered transport related pollution and congestion, and we have a winning model delivered at one's doorstep.

Many cities, some old, some bursting with large populations and some developed haphazardly with ad hoc infrastructure are struggling for space. Moreover parking is a problem and cost of retail spaces can be very high. If you add the cost of stocking and transporting goods, it becomes obvious why this new model of e-commerce which I have dubbed 'K-Commerce' is a compelling alternative. An infinite number of products in infinite configurations and customisations can be delivered to a customer from just small outlets manned by these self-employed youth. Advanced manufacturing and 3D manufacturing is set to revolutionise the retail and e-commerce trade in ways we can only imagine. As you read advances in 3D printing and advanced manufacturing are being made. Once the 'Tipping Point' is reached a new wave of technologies are going to sweep across the world and challenge the established well-oiled supply chains of manufacturers and distribution channels of established retail/e-commerce players.

Conclusion

The e-commerce industry is going to change big time. The change is going to be as big if not bigger than what is going to happen to the oil and auto industries. The changes are going to challenge the way we think and shop. It is going to shake the very foundations of the global trade and also the dynamics of power in the world.

Welcome to the innovative and unbelievable new world of 'K-Commerce'. Your neighbourhood mom-pop store (Kamadhenu) at your service! The Mahatma may well have had the last laugh!

Chapter 11

THE FUTURE OF HEALTHCARE

In the future healthcare is going to be pro-active, pre-emptive, personalised and precise

"When health is absent, wisdom cannot reveal itself, art cannot manifest, strength cannot fight, wealth becomes useless, and intelligence cannot be applied."

– *Herophilus*

"It is not the strongest of the species that survives, nor the most intelligent, but the one most responsive to change."

– *Charles Darwin*

"Although we have enough healthcare support, often it doesn't reach the poor and needy. In this scenario, technology is the best solution."

– *N.R. Narayana Murthy*

Healthcare was and is a sector that is technology centric. Great improvements in healthcare have been brought about over the years with innovations and breakthroughs in healthcare technologies. The challenge, however, is the cost of healthcare and public access to good quality timely healthcare services. Other challenges are those that relate to the personalization of care, the precision of healthcare and the pre-emption of diseases.

These are the areas which the new generation of breakthroughs seek to address. The focus has been on making healthcare accessible, cost-effective, personalized and pro-active. The future of healthcare is going to be nothing short of science fiction coming to reality and there is hope that a quantum leap in healthcare would augur well for the good of all humanity. We are all waiting for healthcare 2.0 with bated breath.

Future of Innovative Healthcare

"In healthcare, we are beginning to see that AI can read the radiology images better than most radiologists."

— Andrew Ng

"The day healthcare can fully embrace AI is the day we have a revolution in terms of cutting costs and improving care."

— Fei-Fei Li

With the coming of age of AI (Artificial Intelligence), big data, robotics, genomics, and material science, breakthrough innovations in providing a more efficient healthcare system at affordable cost are imminent. An example of this can be seen in the field of imaging and diagnostics. This field is resource intensive requiring a large number of highly trained human technicians and doctors across specialties. Technology is bringing about vast changes in this domain with 'Machine Learning' or ML. Machine learning can vastly help in the field of imaging and diagnostics to reduce the workload of specialists.

On the other hand, AR (Augmented Reality) and VR (Virtual Reality) can assist in training doctors and nurses in healthcare. We are likely to see 'Robo-surgeons' and other such innovations in the future. We will also observe more pro-active and personalized healthcare rather than reactive healthcare. Devices could be woven into our clothes to collect data, help in monitoring health and sending alerts as required. Devices could help to monitor heart rate, blood pressure, fluctuations in weight, bio-medical signals from the skin and activity levels of the patient. The treasure trove of data gathered from these devices could help in studies to understand the individual patient's condition and in providing tailor-made healthcare solutions.

Data derived from genomic sequencing could also richly contribute to personalized healthcare. Precision medicine along with personalized and customised healthcare could be used to better treat patients. For instance, the dosage of medicine to be administered to a sick patient could be dynamically altered and prescribed by obtaining the real-time temperature of the patient. One could also decide the course of treatment based on the patient's genetic code and how best he/she would respond to a particular treatment. Going even further we may see custom drugs manufactured for a patient based on his/her genetic makeup and disease.

With the advances in genomics and the ability to handle big data we are entering a phase where personalized healthcare would be standard for everyone. The cost of these technologies has vastly declined, making them more and more mainstream. Along with genomics, wearables can provide data to gain significant insights into how best to treat an individual patient at a personalized level and in a precise manner. Pre-emptive, personalized, precise and tailor-made treatment and medicines are the future of health care.

Future Hospitals

"If we are to ensure that healthcare remains affordable and widely available for future generations, we need to radically rethink how we provide and manage it - in collaboration with key health system partners - and apply the technology that can help achieve these changes."

– Frans van Houten

"The best doctors give the least medicine."

– Benjamin Franklin

Modern hospitals have developed in spurts like patchwork. Improvements in emergency rooms, operating theatres, laboratories, and imaging centres have been built around ad hoc solutions and in a piecemeal manner. This has led to bloated and redundant infrastructure with a need to re-imagine and re-engineer the hospitals from the ground up.

With new digital technologies and as robotics technology matures many mental and physical tasks that doctors perform will be automated. There will be more tele-medicine with lesser need for hospital visits. The future would not just be filled with distance diagnostics and prescriptions but also with tele-surgeries or remote surgeries where doctors from one continent would perform complex surgeries on patients in maybe another continent. We could expect most healthcare delivery to be been done virtually rather than in person. This provides for savings in commute and frees up space in hospitals. In this manner, quality healthcare would probably become more accessible, cheaper and definitely more convenient.

Healthcare is likely to see a 'hub and spoke' model where a number of local clinics would provide outpatient diagnostics and elective care while hospitals would act as hubs providing more advanced care for surgeries and emergency treatments. This would be a more cost-effective model and one with greater reach and accessibility for patients.

Blockchain technologies are going to revolutionize record keeping and the way we access patient's data. Blockchain technologies would give the patients complete control over their medical records. The records would be accessible to each individual from anywhere and any healthcare provider could get access to it with the patient's consent. This would provide the patient with instant, complete and convenient access to his/her medical records. This would also bring in much-needed transparency and would restore the strings of control in the patient's hands.

Future of healthcare would leverage technologies that can monitor information to take action without human involvement. It would provide on-demand services including home or virtual visits and soon waiting rooms may be eliminated.

Incredible Devices and Implants

"Our healthcare system has seen some of the greatest achievements of the human intellect since we started recording history: We're developing incredible devices and implantables to improve the quantity and quality of people's lives."

– Dean Kamen

"The best six doctors anywhere and no one can deny it are sunshine, water, rest, air, exercise, and diet."

– Wayne Fields

In the year 2003, Thomas Boland of Clemson University patented the use of inkjet printers to print cellular constructs. The process involved using a modified spotting system for the deposition of cells on substrates so that they could be organized into 3-Dimensional matrices. Since the invention of bio-printing or 3-D printing of biological structures, we have moved unto production of tissues and even organ structures.

Organ printing has been seen as a potential solution to the problem of organ donor shortages. Artificial organs are often prepared for transplantation by using the recipient's own cells. While currently, there has been success in printing and implanting structures that are either flat such as skin, tubular such as blood vessels or hollow such as the bladder structures, soon bio-printing of more complex organs such as heart, pancreas and kidneys would become mainstream.

3-D (Three Dimensional) printing allows for layer-by-layer construction of organs. The process starts out by first forming a cell scaffold and is followed by a process called cell seeding. In cell seeding, cells of interest are pipetted directly into the scaffold structures. New techniques that preclude the need for cell seeding are also being explored where the cells are integrated into the printable material itself.

Bio-printing uses modified inkjet printers to produce 3-D biological tissue. Here smart gel and a suspension of living cells are used in the printer cartridges. By printing the smart gel and living cells alternatively, the cells eventually fuse together to form a tissue with the smart gel providing for structure. Finally, the gel is washed away leaving only the living cells. In this way, tissues and organs could be created.

3-D (Three Dimensional) printing is making waves in dentistry also. A complete set of 3-D dentures with gum and teeth have been printed successfully using biocompatible material. Considering that even the 3-D denture base is 3-D printable, it can be adapted to a patient's morphology. Similar to 3-D printing of dentures, we can now even print 3-D crowns, aligners, and bridges. It is possible to scan patient's teeth in all three dimensions and create a model. We can then print a 3-D crown based on it. Not only does it save time, but the crown or denture can be reprinted if there is a mistake. We could even print night guards, surgical devices and tools which are perfectly adapted to surgeon's working methods or processes.

In the future, light activated 'bio-glues' would be used instead of stitches. These are not only non-toxic but also adhere to even wet biological tissue surfaces within seconds. They can even resist pressures like that of a beating heart. They are composed of a gel made up of proteins and other molecules and require ultra-violet light to activate and adhere. While currently these have been successfully tried on pigs, where they have sealed carotid arteries, soon we will find them approved to be used for humans.

The field of robotics has taken a new turn with the concept of 'Exoskeletons'. These are assistive technologies to help people with 'Activities of Daily Living' (ADLs) such as mobility and communications. These exoskeletons are custom made contraptions that act as supporting or enabling assistance devices that could, for example, allow paralyzed people to walk using their mind as a control center. These could also help individuals who are differently abled or individuals with age-related mobility issues to go about their daily lives in a non-dependent way.

Robotics also plays a part in healthcare where proper robots with AI could act as companions for children with mental disabilities or provide assistance or company for older people. Robots could also understand if there is an emergency and call the ambulance or alert the right doctor when necessary.

Wider research in the field of brain-computer interfaces or brain implants has shown much promise. These neural implants as they are called have been used in deep brain stimulation and Vagus nerve stimulation. These simulations have been found to be effective in the treatment of Parkinson's disease and clinical depression. Research has also been ongoing in an attempt to create a biomedical prosthesis that would circumvent stroke affected or injured dysfunctional areas in the brain. It is hoped that one day the use of brain implants would help restore vision, hearing, movement and speech in stroke patients.

Data As a Consumer Good

"An ounce of prevention is worth a pound of cure."

– Anonymous

"Take care of your mind, your body will thank you. Take care of your body, your mind will thank you."

– Debbie Hampton

As incredible as it may seem, in the future, one may actually be paid to be cured. Much of the healthcare in the future would be done on the back of consumer data. When one contributes to it, one may actually be paid for it. Individual consumers may be able to monetize their data set by granting access to the healthcare ecosystem of their personal data. They could derive benefits by providing access. Organizations receiving the data set would be able to enhance their services and get advanced scientific insights into health conditions and help to further science in this manner.

Imagine a world where your personal health data could be "currency" to pay for your treatment. It would be a world where devices constantly monitor your health even as you go about your daily tasks. The demand for hospital beds and the frequency of patient visits to the doctor could be efficiently managed. The patient is tended to when his/her physical parameters stray. The patient is observed, data is captured and the patient would be treated without need for human intervention. Medicines in the right dosages may be delivered to the patient based on real-time monitored data and analysis.

Interestingly, signs of this future can be seen in Amazon's *Alexa*, a voice enabled assisting device. Amazon has patented a technology where Alexa knows when you have a sore throat and asks you if you want to book a doctor's appointment or get a virtual consultation. After the initial virtual consultation, it may send you a test kit or portable device to do some tests for things like a strep throat. It may then send you antibiotics based on the results of the test. All this is done within hours before you even leave your home.

It is estimated that by 2020 there would be such a flood of medical data and progress that it would take just 73 days for all the medical knowledge to double.

In the year 1950, it was estimated it would take 50 years for the same to happen. AI would play a prominent role in helping doctors keep up with this huge trove of medical information as more and more medical data, diseases, conditions, treatments, medical technology and research comes online. The interoperability of all the medical data and information from around the world would make it a huge trove that would benefit mankind in a big way. It is estimated that a single individual in his/her lifetime could generate data that could fill hundreds of millions of books. All this data needs to be analysed, it needs to be comprehensively collated. Big-Data and AI are going to play a major role in this.

In addition to these revolutionary changes one's personal health data may also help to cut down on one's insurance premiums. Those who take care of their health by exercising and keeping their health parameters in range may be rewarded with lighter premiums. This would encourage them to stick to their disciplined ways. One's personal health data may also be used to help with preventive health care to pre-empt conditions. For e.g. sensors in your wrist and socks may be able to tell a lot about your postural sway and likelihood of falling. This, in turn, could be attended to, to prevent an untoward fall and injury. This would help to pre-empt the chances of an accidental fall in the case of senior citizens. Also, broken bones in older people and the resultant hospitalization from falls could be avoided; thus saving money and a lot of pain and agony.

Similarly, a new technology that is in use incorporates sensors that are woven into the individual's garment/jacket. These sensors measure ECG/EKG's (Electro Cardio Gram) and may be able to predict a heart attack and help to pre-empt it with intervention. Just as data can be used to pre-empt conditions, the data on the efficacy of the treatment given and post data analysis can yield significant insights. These insights act as a feedback loop enabling the constant and continuous improvement of treatment methods.

Food is thy Medicine

"When diet is wrong, medicine is of no use. When diet is correct, medicine is of no need."

– Proverb from Ayurveda

"Let food be thy medicine and medicine be thy food."

– Hippocrates

"Those who think they have no time for healthy eating… will sooner or later have to find time for illness."

– Edward Stanley

The future of healthcare would lay emphasis on prevention. As the saying goes, "a stitch in time saves nine". It is common knowledge that what we consume on a daily basis lays the foundation for good health or disease as the case may be. While genetics do play a role in how healthy we are, our lifestyle choices would also have an effect on one's health. Prevention has to start from what we eat rather than providing band-aid solutions after problems become entrenched. What Michael Pollan once said applies here also. He said, "If it came from a plant, eat it. If it was made in a plant, then don't."

Future healthcare would incorporate recommended diets based on one's level of activity, climate, geography and season. Based on sensors which one wore, the daily requirements for one's health would be calibrated. For example, one could need extra calcium based on one's active lifestyle or may need more folic acid during the flu season. Our diet would be automatically calibrated and we may be sent the season's fruits and vegetables automatically through a grocery delivery e-commerce company along with great recipes.

Interestingly, the ancient science of Ayurveda from India treats food as medicine. It prescribes diet and cooking that takes into consideration the effect of the ingredients, spices and herbs on one's health or symptoms. Diets and recipes are recommended based on one's health problem. Food is part of the cure and prevention. Various plants and herbs are concocted and given to patients to enhance the ability of the body to fight the disease holistically. This helps to solve problems from ground-up. Hence it targets the root cause of the problem. Rather than just suppressing the symptoms it treats the imbalance that caused the symptom. Rather than ingesting quick-fix pills, the healing takes places over weeks as the body's circulatory system absorbs ingredients and heals. We are going to see more such holistic healing and treatments in the future as practitioners of healthcare begin to treat the body as a holistic system rather than disparate parts…

Conclusion

"The secret of health for both mind and body is not to mourn for the past, not to worry about the future, or not to anticipate troubles, but to live in the present moment wisely and earnestly."

– Buddha

"Happiness is nothing more than good health and a bad memory."

– Albert Schweitzer

It goes without saying that health is central to the happiness of a human being. Being in good health is a gift we need to cherish every single day of our life. Healthcare of the future would be easily accessible at our fingertips and affordable. The healthcare would be affordable as the benefits of technology get spread over a large swathe of people around the world. While improvements could be incremental, in many instances it would be revolutionary with genomics and new sensors. Nano health, genomics, custom medication, brain implants, artificial organs, networked sensors, exoskeletons, PPP (Preemptive-Precise-Personalised) healthcare, etc. are just some of the buzz words we are going to hear more of in the future.

It suffices to say that healthcare in the future will be a lot different from what we have now. It would be more data-driven and personalized with emphasis on precision and pre-emption. The importance of health cannot be overemphasized. Without health, no human progress is worth it. In this context, technology and systems of the future would play a huge role in enhancing every individual life on the planet. I would conclude by leaving you with an Arabian proverb on the importance of health.

"He who has health has hope, and he who has hope has everything."

– Arabian proverb

Chapter 12

AGEING AND DISEASE

Why you are going to think differently about ageing and disease

"Rather than looking at 'Ageing' as inevitable we need to see it as a condition that can be treated and prevented. It is a change in perspective that is the need of the hour."

– Author

"Ageing is NOT a given or a one way process. It is plastic and can be subject to intervention or even reversal."

– Author

From the 1900s to 2000s the life expectancy of a human on this planet has gone up by about 30 years, from 35–45 years to 65–75 years, with variations depending on race and geography. Apart from better healthcare and prolonged life of the elderly, a major contributing factor has been the fight against infant mortality. However, from the 2000s onwards, life expectancy is expected to double in the future.

The list of diseases that have been conquered in the last one hundred years includes – polio, typhoid, tetanus, yellow fever, measles, black plague, diphtheria, chicken pox and smallpox. Many of which have almost been eliminated in most parts of the world. We have fought back with vaccines and powerful drugs to save children and adults alike from the scourge of early death.

The Current Challenge of 'Lifestyle' and 'Age' Related Diseases

"Ageing is, simply and clearly, the accumulation of damage in the body. That's all that ageing is."

– *Aubrey de Grey*

Though we may have fought and may be even won a few of our battles against most diseases that ran amok through many centuries, we have now come to a point where our greatest hazards have become two-fold. The twin challenges are the threat of lifestyle diseases and the ever present conditions brought about by ageing. While lifestyle diseases can be managed by an adjustment to lifestyle and are mainly a problem brought about by modern conveniences, ageing has been a problem that we have been unable to shake off.

As we age our cells begin to wear off and sometimes malfunction or even stop working. This leads to conditions like heart disease, arthritis, Alzheimer's disease, cancers, Parkinson's disease, diabetes, hypertension and many other age related problems that crop up and stifle our ability to live life to our fullest potential. Some researchers believe that rather than treating ageing as ineluctable we should maybe start looking at it as something that can be prevented and treated. Recent science has pointers to show that biological ageing maybe entirely preventable and even treatable.

The body ages according to genetic and environmental factors that shape its rate of deterioration. As minuscule errors begin to build up in our DNA, our cells begin to develop faults which culminate in tissue damage. This accumulation and culmination result in slow deterioration of a healthy body which begins to get afflicted by chronic illnesses.

While it may not be possible to stop ageing altogether there is much hope to slow ageing and this may potentially give people an extra 30–50 years of life. It may also be possible to prolong youth. It is even more valuable to give the person the ability to stay young, fit and healthy for a longer duration than just increasing lifespan. For instance, it was shown that one common diabetic drug called metformin was able to extend the lifespan of rodents. Even in the early 1990s, Cynthia Kenyon, who is now the vice president at Calico Labs (which is a Google backed anti-ageing company) demonstrated that roundworms could double their average lifespan of three weeks by just changing a single letter of their genetic code. These so called 'rejuvenation technologies', can be used to revert the cells of older people to what it was in their youth. A 60-year-old may be reverted to biologically feel like he/she is 30 years old. These however, are not permanent fixes and ultimately can only delay the inevitable.

In 2013, a study by researchers at the Harvard Stem Cell Institute showed that the muscle strength of mice could be improved by a growth factor found in the blood of younger ones called GDF 11. Though these studies could not be replicated it does point to possibilities. Among one such spin-off method is something called the 'Vampire therapy'. When dementia patients were infused with blood plasma from younger donors between 18 and 30 they showed signs of improvement. Even early onset Alzheimer patients, regained their ability to bathe and dress themselves or do other household tasks. Some start-ups have seized this opportunity and are offering customers blood transfusion from donors between 16 and 25 years of age, at a price.

Whatever may be the case, it is an established fact that ageing is not given or a one-way process. It is in fact plastic, can be subject to intervention and or even possible reversal. Ageing is the foundation of many conditions such as cancer, Alzheimer's disease, Parkinson's disease, dementia, etc. These conditions, in turn, can be tackled by tackling ageing. Thus, it may be possible to kill many birds with one stone.

P-Medicine

"Ageing's alright, better than the alternative, which is not being here."

– George H.W. Bush

The future of precisely tailored health recommendations (according to genetic type) will be accelerated due to recent phenomenal advances in genetic research combined with the current capability in high-performance computing necessary for the analysis of large datasets. The science called *'Genomics'* has begun to take a shape of its own with vast applicability and accessibility in a relatively short span of time. What really started out as a discovery of genetic mutations caused by diseases like diabetes and cancer has now moved on to treatments that are personalized based on an individual's genome. A range of tools that include third-party cloud for data sharing and artificial intelligence have enabled and have accelerated this shift. Nowadays, software and powerful computing can be used to identify patterns in extremely large data sets. This provides pointers to potential links between specific genes and proteins, to diseases. This could then be tackled at an individualized level. This would be a more effective approach and is termed as *'Precision Medicine' or 'P-Medicine'*.

The 'P-Medicine' as it is called is more personalized, precise, preventive, predictive, pharmo-therapeutic and participatory. The objective of such a type of approach is to deliver to the right patient, the right kind of behavioural prompts, at the right time to maximize good health.

Nature vs. Nurture

"According to the famous psychologist Sigmund Freud, every person has three levels of personality. And the third one is our superego... This is our conscience. The superego is directly influenced not only from our internal mechanism but also by the mores or rules of society."

– Carol Roach, Sigmund Freud and the super ego

Over the years, the cost of genetically sequencing a human has dropped dramatically from a few million dollars to a few hundred thus increasing accessibility and applicability of genome sciences. It would even be possible for one to send a sample of one's saliva via mail and receive a report for the same. Another field that is gaining prominence is the field of *'Nutrigenomics'* which studies how dietary factors interact with one's genes. Since people have different genes, ultimately there is going to be a variety of solutions for different people. By knowing about one's genetic make-up and adjusting one's diet one may be able to dodge a number of lifestyle and age-related diseases.

Another important factor that affects our age-related progression is related to early exposures one may have had to elements in one's immediate environment. It has a major impact on the manner in which our genetic potential actualizes itself. If one could be told in advance what the early life optimal conditions that favour the good sides of our gene are, then one could get a head start and make those early adjustments. This field called *'epigenetics'* tells us the effect of 'nature and nurture'. This would allow us to make the required adjustments early on and perhaps also stall the early onset of age-related diseases.

A physician could review your lifestyle, sleep habits and health history. The physician would also get some blood work done to compare biomarkers with baseline measures from your twenties. He/She would then give you a prescription for the changes you need to adopt. This would include a diet that mimics the effect of fasting and a drug that helps your cells clear out any malfunctioning proteins with the goal of lengthening your lifespan and slowing your ageing process. *'GeroScience'* as it is called explores the relationship between ageing and the onset of diseases like cancer and heart disease. The idea is that there is a fundamental shift in managing age-related diseases where the entire group of diseases could be tackled by only tackling the process of ageing alone.

Key Traits Associated with Ageing

"The great thing about ageing is that your eyesight deteriorates at the same rate as your face. So I can't see how bad things are getting!"

– Sue Perkins

An important step in this approach is to define what constitutes ageing. These are key traits or hallmarks. The field of such study is called *'biogerontology'* and lays down 9 traits associated with ageing. They include the following

- **Genomic Instability**: This refers to both environmental and genetic factors that cause damage to one's genes as one goes through one's life. This accumulated damage accelerates the pace of ageing.
- **Telomere attrition:** *'Telomeres'* are protective caps that are located at the ends of our chromosomes that house our genetic material. These start getting shorter each time a cell divides. Finally, there comes a point that the cell is unable to divide anymore which can lead to disease.

- **Loss of Proteostasis**: When cellular proteins become misfolded due to ageing and therefore lose their homeostatic functions, it causes the build-up of damaged proteins and is a sign of ageing.
- **Epigenetic alterations**: These are changes to the expression of genes due to an individual's life experiences and other environmental factors affecting ageing. (These are changes to the expression of genes and not just changes to the DNA, is something to be noted).
- **Cellular Senescence**: When older cells cannot be cleared out as fast, their build up are biomarkers for the ageing process.
- **Mitochondrial dysfunction**: When the *Mitochondria*, which is the energy powerhouse for regulating metabolism starts to malfunction with age we begin to feel a slowdown.
- **Stem Cell Exhaustion**: There are 4 types of stem cells which help in regenerating tissue. Their decline is one of the biomarkers of ageing.
- **Altered intercellular communication**: When intercellular communication between cells is disrupted with age, it can cause inflammation and age-related diseases to crop up.
- **Deregulated nutrient-sensing**: When metabolism-regulating pathways whose proteins are influenced by nutrient levels are affected it can promote ageing.

There are a number of companies tackling one or more of these biomarkers with varying degrees of success.

One prominent approach focuses on *'Cellular Senescence'*. While in normal human beings the cells keep dividing throughout one's lifetime at some point they may stop dividing and keep accumulating without being eliminated. These 'Zombie cells' keep moving around in our bodies and may lead to inflammation and age-related disease. The research focused on targeting and eliminating these 'Zombie cells' have been in play for a while.

One popular area of research is around *'Senolytics'* which is a class of drugs that target ageing or senescent cells and destroys them via induced cell death. The 'Senotherapy' as it is called is a new area of research that is able to prevent or reverse ageing by targeting cellular triggers such as damage to DNA cells.

Other focus areas include the telomeres and prevention of their degradation. Research has also focused on the energy powerhouse of our body the Mitochondria and also on *mTOR* which is a regulator of cell growth and proliferation.

In the year 2016, Japanese researcher Yoshinori Ohsumi was awarded the Nobel Prize for medicine for deducing the mechanisms of *'Autophagy'*. Autophagy allows for unneeded proteins to be degraded and also aids recycling of amino acids. This helps in synthesis of proteins essential for survival. By practising intermittent fasting, exercise and diet one would be able to incorporate 'Autophagy' into one's life. This helps to keep one younger and lessens probability of being afflicted by cancer, Parkinson's or Alzheimer's diseases. Pharmaceutical companies around the world are trying to research and produce drugs that mimic 'Autophagy' and deliver the benefits associated with it to patients.

Regenerative Medicine

"For healthy adult people, the really big thing we can foresee, are ways of intervening in the ageing process, either by slowing or reversing it."

– Nick Bostrom

'Regenerative Medicine (RM)' has four approaches to solving the age-related issues namely:

- **Cell Therapy**: This involves using living cells to replace or augment diseased or damaged cells.
- **Gene Therapy**: This involves altering the genetic code to treat the diseased state.
- **Tissue Engineering**: This involves the production of replacements for tissues or organs including the use of bio-materials, synthetic materials and 3-D printing of tissues and organs.
- **Small Molecules and Biologics**: This involves the stimulation of dormant cells for regenerative purposes.

Combinations of regenerative therapy such as (cell therapy + gene therapy) etc. are also gaining momentum. A combination of a number of gene-editing tools, have been used widely to achieve these objectives.

If 'Life is Short, Extend it'

"We need a new and positive narrative that defies the negativity and transforms our ageing populations from a problem into an asset."

– Alexander De Croo

"Aging followed by death is the price we pay for the immortality of our genes. You find this information soul-killing; I find it thrilling, liberating."

– David Shields

One of the first anti-ageing measures involves 'Managing' the process of ageing. For example, one may use dietary supplements to extend one's general health and wellness. Next, one may take steps to 'Repair' age-related diseases. By repairing one of these problems, one may seek to treat the age-related problems. For instance, one may use 'Regenerative Medicine' to treat age-related problems. The inter-relatedness between age and related diseases, or age and age revealing changes, makes it possible for research to find solutions for both, while finding solutions for one. While studies and experiments are directed towards improving aging skin by removing wrinkles and fine lines, this may be ancillary with the larger intent and focus directed on 'preventing' or maybe even 'reversing' the ageing process. This is something that is cutting edge technology that is still in development. It offers hope for the future where one would be able to delay and reverse ageing at least temporarily such that one leads a healthier life with a longer lifespan. A day will dawn when being 100 would still be termed as middle age and 60 would be one's youth. People would normally live to be over 150. This has immense ramifications for society where one would get to see one's great-great-grandchildren and live a full and complete life.

Elixir of Immortality

"Physical immortality is seductive. The ancient Hindus sought it; the Greek physician Galen from the 2nd Century A.D. and the Arabic philosopher/physician Avicenna from the 11th Century A.D. believed in it."

– S. Jay Olshansky

"Millions long for immortality who don't know what to do with themselves on a rainy Sunday afternoon."

– Susan Ertz

While extending one's lifespan seems within reach and is impressive, what about taking on death and drinking the elixir of immortality? For a long time, there have been proposals to cryogenically freeze a person's brain and body immediately after death. The intention was to revive it at some later point, when technology would be sufficiently advanced to bring it to life. A number of companies have been offering this to wealthy clients as a way to preserve their bodies. One such company is *'The Alcor Life Extension'*. However, while the future may still hold the promise, as of date, no client of the company has been resurrected from their icy cryogenically frozen storage units.

While preserving the body is fine; you may ask what about the brain? Once the body dies the circuits in the brain may just blackout for lack of oxygen. How was one supposed to revive that? Such questions are logical and justified. Futurists, such as Ray Kurzweil who is also the theorist of the *'Singularity'* theory and lead engineer at Google, has espoused 'mind uploading' as a way to at least achieve digital immortality.

Conclusion

In the future, with advances in medicine, sixty years old would still be considered youth. Senior citizens would be a classifications reserved only for the centurions. Advances in medicine would not only give one long life, but also extended youth and vitality. This would allow one to life a full life and achieve all of one's life goals.

Beyond that, humans may make the moon-shot for immortality. You never know, you may well live into the future with your body in an icy storage unit and your mind uploaded into a hard disk. The future sure seems different from being six feet under!

Chapter 13

NANOTECHNOLOGY - SMALL IS BEAUTIFUL

As we begin to miniaturise things we also make them exponentially more powerful

"Sometimes beautiful things come in small packages."

– Author

"Strength, durability, power, efficiency, flexibility, conductivity and many other characteristics can be vastly improved by paying attention to the small things at a nano-scale."

– Author

I was once on the phone chatting with a chip design engineer more than a decade and a half back. As we chatted I began to ask him questions related to chip design. I began to learn a lot about the field. Famously there was *Moore's law* which was an observation of a historical trend by Intel co-founder Gordon Moore who stated that the number of transistors on a chip doubled every two years (in practice it was even quicker and it was more like 18 months).

The one thing I gleaned from the conversation was that the smaller the electronics, the better, faster and more energy efficient the electronic devices. Electronic functions occur when communication within the electronic device is achieved with the movement of electrical signals within the circuits. As the length of the connecting path is reduced, the system is more energy efficient.

Thus smaller the length of the path traversed, the lesser the loss of energy in the form of heat and greater is the efficiency of the functioning circuits. This caused a quantum leap in the performance of the chips. Of course, there was much more to the science and physics of it, but this gave me a glimpse into why sometimes smaller is better and more beautiful.

When *nano-technology* began to take hold, it got me really excited because of the infinite potential of what could become possible. What kick-started the nano-technology thought process in the scientific world was a talk titled "There's plenty of Room at the Bottom" by physicist Richard Feynman, at the American Physical Society meeting at Caltech, at the end of the decade of the 1950s. Much later, the term nano-technology was coined by Professor Norino Taniguichi. However, it wasn't until even much later in 1981 with the development of the scanning tunnelling microscope when one could see individual atoms, that modern nano-technology began.

To really imagine what nano really is, is really hard. About 100,000 nano-meters are about as thick as a newspaper. For instance, if we consider a marble to be a nano-meter, then a meter would be the equivalent of the size of the earth. Nano-technology involves the ability to see and control things at a molecular or atomic level. Nano-technology presents advantages of enhancing the properties of matter to give higher strength, lighter weight, greater chemical reactivity and even control of the light spectrum.

Broad Applications of Nanotechnology

Nano-technology has wide-ranging applications. 'Smart fabrics' with nano-sensors are used to monitor health, nano based solar cells are used to capture solar energy more efficiently, nano-technology is even used to make lightweight cars, trucks and even spacecraft. They are everywhere.

Nano-films have been used on eyeglasses, computer, mobile, and camera displays. This would make them water resistant, anti-reflective, self-cleaning, anti-fog, anti-microbial, scratch resistant and electrically conductive. Nano-technology has helped in making better sports racquets, bicycles, luggage, helmets and power-tools to make them lighter, stronger, durable and more resilient. They have also been used in body armour to provide for lightweight ballistic energy deflection.

The technology has also been utilised in re-chargeable battery powered systems, low-cost sensors, solar power panels, fuel additives (for cleaner fuel), in lubricants (to reduce wear and tear) and even in aircraft fuel storage tanks (for providing improved thermal, mechanical and barrier properties). They are also widely applied in petroleum refining and in automotive catalytic converters.

They work even in household products to remove stains, in filters and in air purifiers. They are also used in stain-resistant paints. Nano-scale ingredients such as titanium dioxide and zinc oxide have also been useful in sunscreens providing for UV protection from the sun.

Nano-Technology in Infrastructure

The nano-engineering of construction material like aluminium, steel, asphalt, concrete offers great promise in terms of improving longevity, sturdiness, performance while reducing life-cycle costs. New innovations like self-repairing structures and the ability to transmit energy may provide new dimensions to possibilities.

Continuous monitoring of the structural integrity and conduct of bridges, tunnels, rails, parking structures, and pavements is made possible by this technology in a cost-effective manner. Nano-scale sensors may aid in transportation where it would facilitate communication with drivers to maintain lane discipline, adjust travel routes, avoid congestion and importantly avoid collision.

Revolutionizing the Energy Sector

Nano-structured solar panels would be cheaper and more effectual. As films they are easier to manufacture and install, in the future they could even just be painted on different surfaces. Wires containing nanotubes may be used in the electricity grids providing for better conductivity, lower heat losses, and more effectual power grids. Quick charging batteries that are lighter, more efficient with greater power densities are also being developed. Epoxy coatings with carbon nanotubes are making windmill blades long lasting, stronger and lightweight. This has helped in generating more electricity per revolution.

Piezo-electric nanowires interwoven with clothing and thin film solar could power personal electronic devices into the future. Nano-technology can also help in ethanol production from corn, wood chips and grasses. Cellulose-based nano-products can be used in a wide array of sectors including packaging, food, healthcare and energy.

Nano-Technology is used in producing the catalyst used in petroleum refining and also helps to reduce fuel consumption in vehicles and power plants, contributing to higher efficiency through better combustion and decreased friction.

The technology has also been deployed in oil and gas extraction industries where they are used to detect fractures and leaks. Nano-tube scrubbers are useful to make the power plant exhausts less polluting.

In space, nano-technology could be a game changer. By the development and use of advanced nano-materials which have twice the strength of conventional composites at a fraction of the weight, it is estimated that the weight of a launch vehicle could be reduced by two thirds. This not only reduces the amount of energy to launch a vehicle but it would also make the development of a single stage earth to orbit launch vehicle possible.

Revolutionizing Computing and Technology

Nano-technology has been revolutionizing computing for long. The day is not far off, when the entire memory of a computer could be stored on a tiny single chip and that would be just the beginning!

Instant boot, instant save computers, ultra high definition displays and television with the quantum dot technology that could display even better, are technologies that would transform the retail markets and be available soon at the nearby store.

Electronics would be more energy efficient. They would come in configurations that are bendable, foldable, stretchable, rollable and potentially cheaper with wide-ranging applications. From wearables to IoT (Internet of Things), a revolution has just been seeded. These devices would come in the form of tattoo sensors, miniature photovoltaic cells rolled in clothing, paving way for countless applications in wearables, medical application, communication and assistive technologies.

In the future, we can expect flexible display e-readers, conductive inks, RFID enabled packaging, more powerful mobiles, thumb drives and even powerful hearing aids.

Nano-Science in Medicine

Nano-science opens a whole world of applications in the treatment of cancer and other diseases. Nano-Science has potential to treat lifestyle diseases like cardiovascular complications arising due to the build-up of plaque in arteries. The application of nano-technology to fight cancer has a number of benefits including targeted delivery of drugs to cancer cells to minimize damage to healthy tissues.

Nano-technology could play an important role in regenerative medicine in neural, bone and dental application. The technology could be used to grow human organs and even repair spinal cord injuries with *Graphene* nano-ribbons. Also, there has been promising research to show that neurons grow well on conductive Graphene surfaces.

Nano-medical research has also been looking at ways for better vaccine delivery that would be without needles. There is also on-going research for efficient and quick response development of vaccines for new threats. Nano-technology has also helped in developing better imaging and diagnostic tools which pave the way for early diagnosis and individualized treatment for better therapy. The technology also helps in genetic studies with high speed and low-cost gene sequencing technologies.

Nano-Technology and the Environment

Nano-technology helps in quick and low-cost detection of impurities and its treatment, thus laying the foundations for providing clean and affordable drinking water to the masses. Desalination of seawater using thin films with nano-pores made of molybdenum disulphide (MoS_2) has the potential to filter larger quantities of water making it better than current systems.

Nano-technology has its applications in pollution control where the cleansing of industrial pollutants using nanoparticles that produce chemical reactions to render the pollutants harmless has been developed. Magnetic water-repellent nano-particles have been used to remove oil from oil spills.

Nano-technology based air filters that utilize fibre material with nano-pores to trap particles and particulate matter have also been developed. These are widely used in modern airplanes and offices. Nano-technology functions within sensors to detect and identify chemical or biological agents. Fire-fighters, for instance, can use a sensor attached to a smartphone that can detect deterioration in air-quality around fires and allow the fire-fighters to make an exit before things turn bad.

Graphene

A special mention has to be made on *Graphene* which promises to revolutionize and open a new dimension to herald infinite possibilities in the field of nano-technology itself. In the year 2004, two professors from the University of Manchester discovered and isolated a single atomic layer of carbon for the first time. This breakthrough won them the Nobel Prize in Physics for 2010.

The professors used a *'Scotch tape method'* also called as a *'Micromechanical cleavage technique'* which was so simple and effective that hundreds of laboratories around the world began following that method for their Graphene research without the need for expensive equipment.

Graphene is a honeycomb-like pattern of one atom thick sheet of carbon. It is the world's strongest, thinnest and most conductive material with the potential to revolutionize sectors such as energy, battery, sensors, electronics, and appliances.

The vast applications of Graphene include Graphene enhanced composite material which, because of its strength and its lightweight quality, finds applications in building material, aerospace, mobile devices and many others.

Graphene finds applications in energy storage such as fuel cells because it is the world's thinnest material with the best capacity to conduct electricity. This allows it to have the highest surface areas to volume ratio making it ideal for batteries and super-capacitors. Graphene enhances the energy density of such products and also allows for faster charging.

Graphene is also used in anti-corrosion coatings and paints, in electronics, flexible displays, drug delivery, sensors, DNA sequencing, health-monitoring wearables, camera sensors, next generation spectrometers, clean water applications and many more. The future of Graphene sure looks promising from here.

Conclusion

Nano-technology in its own 'SMALL' way has the ability to make a BIG difference to our lives and will continue to surprise us with what it can do. The wide-ranging impact this technology will have on our future is something we are only beginning to discover. Sometimes beautiful things do come in 'small packages'!

Chapter 14

ARTIFICIAL INTELLIGENCE

Will humankind meet its match?

"Artificial intelligence is growing up fast, as are robots whose facial expressions can elicit empathy and make your mirror neurons quiver."

– Diane Ackerman

"A year spent in artificial intelligence is enough to make one believe in God."

– Alan Perlis

Years ago, during my Master's in engineering in the late 90s, I was listening to a professor who was giving an introductory lesson on the topic of *artificial intelligence*. He said, imagine that you show a dog to a computer (with image capture, processing, memory and other paraphernalia) and define it stating "This is a dog" and you go on to show it another picture of another dog and again state "This is a dog" and so on. Imagine you do this for some time showing it different breeds of dogs guiding it through a learning process. Then, you finally show it one picture of another animal such as a wolf along with another picture of a dog that it has not seen before. The computer would correctly point out the dog, as a dog and the wolf, as NOT being a dog.

Basically, the computer learns to recognize how to identify a dog by being shown a large enough dataset and then applies the learning to correctly identify a new picture of the same animal, in this case a dog. The system essentially learns to identify a dog after being trained to do the same.

Artificial intelligence was founded on the premise that human intelligence can be so precisely described and captured that a machine could simulate it independent of any guidance.

Turing Test

Alan Turing had described a test for a machine to be considered intelligent. This test is the benchmark that most AI developers seek to attain. Turing proposed that if a human interacting with a machine on one hand, and a human on the other fails to distinguish between the behaviour and responses of the machine from the human, then the machine could be considered intelligent in a human sense.

A converse derivative of the Turing test is what is commonly referred to as CAPTCHA. This is something that most readers would be familiar with. CAPTCHA stands for 'Completely Automated Public Turing tests to tell Computers and Humans Apart'. As described by the expansion, this is a test that helps to determine if the user is actually a human and not a machine. In other words, a computer posing as a human would fail such a test. This is in direct contrast to a Turing test which seeks to determine if the system is indistinguishable from a human. Here, the CAPTCHA seeks to establish that the user **is** human. In the CAPTCHA test the user is to complete a simple test which supposedly a computer cannot do successfully. If the user passes the test then the user is not a machine and it is more likely that the user is a human. A common type of CAPTCHA involves typing of alphanumeric or symbols in a distorted or convoluted abnormal font that appear as an image that a computer cannot decipher.

Capturing the Imagination

"Artificial intelligence would be the ultimate version of Google. It would be the ultimate search engine that would understand everything on the web. It would understand exactly what you wanted, and it would give you the right thing. We're nowhere near doing that now. However, we can get incrementally closer to that and that is basically what we work on."

– Larry Page

"We must address, individually and collectively, moral and ethical issues raised by cutting-edge research in artificial intelligence and biotechnology, which will enable significant life extension, designer babies and memory extraction."

– *Klaus Schwab*

The great increase in computational power of the computer made it possible for AI (Artificial intelligence) to become discernibly perceptive. The increase in computational power of machines has been in tune with *Moore's law* and AI began to be used in a number of fields such as data mining, medical diagnostics, gaming and many others.

'Deep Blue' became the first computer chess-playing machine to defeat a reigning world chess champion, Garry Kasparov on the 11th of May 1997. This captured the popular imagination of the public at large and demonstrated the power of AI. In the year 2011, IBM Watson's question answering machine system played against and defeated comprehensively two of the greatest Jeopardy champions Brad Rutter and Ken Jennings.

Since then AI has evolved and a wide variety of applications of this growing technology have been demonstrated.

Some examples of current and future applications are as follows:

Finance and Accounting

AI systems can take over dull, redundant and repetitive tasks that will free human finance and accounting professionals to do higher level analysis and consultation for clients. According to the well-known consulting firm Accenture, "Automation, mini-bots, machine learning and adaptive intelligence are becoming part of finance teams at lightning speeds".

These AI systems are not only capable of handling more clients but are also built to deliver more value by their ability to determine actionable insights from machine learning and in proposing suitable solutions.

In accounting 'accounts payable and receivable processing' can be automated by an AI-enabled invoice management system that can learn appropriate accounting codes and apply them to each invoice, thus streamlining and automating the digital workflow. 'Supplier on-boarding' process can be automated where the AI enabled system automatically checks for their tax information and credit score in the vetting process. 'Audits' can be digitized and digital trails can be pursued to make the process more efficient, accurate and complete. With the help of AI, 100% audit of companies can be carried out instead of just a sample. 'Monthly information report/quarterly reports' can be generated automatically and quickly, where data can be sourced, consolidated and reconciled with an AI-powered system driving it. Similarly 'expense' reports can be reviewed and approved in compliance with the company's policies. Machines can process a large amount of data, spot patterns and learn to treat the data according to a set of rules freeing professionals to do higher order tasks.

In the financial and banking world, AI-based systems can also help in fraud protection, which would help make investors make safe investment decisions. Machines have the ability to crunch large amounts of data, spot patterns from the past, and present data in distilled form. They also allow for predictions about patterns that may repeat in the future.

Ultra-high frequency trading is one such area, where AI-powered systems are beginning to leave their mark. AI has greatly enhanced 'Algorithmic trading' whose defining pillars have been, limited human intervention and speed. A number of funds, banks and proprietary trading firms have entire portfolios that are managed in whole by 'expert systems' or AI.

Customer Service

AI chatbots can be efficiently used to solve common questions or queries from customers. In banking, chatbots can respond to queries from customers including latest account balances, when bills or payments are due and status of customer accounts.

Machine learning and AI are used to build intelligent conversational chatbots and voice-enabled assistants with conversational interfaces that can answer most FAQs. This has applications in travel, hospitality, concierge, e-commerce, telecom and many such industries which are customer facing.

Music

In the year 2012, the first complete classical album fully composed by a computer was created. AI has even been used to produce computer-generated music for stress and pain relief. To some extent, AI has enabled computers to be able to create human-like compositions.

The evolving field of 'Algorithmic Computer Music' has taken the world by storm. *Sony CSL research labs* has demonstrated the ability of AI to create pop songs by learning music styles from databases of songs by analysing a combination of styles.

IBM has gone a step further. Using 'Reinforcement Learning and Deep Belief Networks', *'IBM Watson Beat'* has created music with an input of a simple seed and a selected style. Since the program is open source many musicians have been collaborating with 'IBM Watson Beat' to compose music.

Security

Security is an area where AI plays a major role. AI enabled security cameras are used to analyse feeds and can prevent crime before they occur. These unmanned cameras alert the security to any suspicious behaviour and allow for timely action. Akin to the movie *'Minority Report'*, AI enabled cameras are being used to prevent shoplifters from shoplifting by spotting suspicious behaviour, alerting security and salespeople to approach the person and ask if he/she needs help, often precluding illegal acts.

With advances in object recognition and facial recognition, security cameras can be used for surveillance and crime prevention on a larger format. Cameras would no longer need to be manned by humans. For instance, once traffic cameras detect a violation, it could even send a parking ticket to an individual on his/her mobile or directly to his/her address without any human intervention.

Agriculture

In Agriculture, farmers are beginning to use automation and robotics to increase the efficiency of de-weeding, harvesting and crop planting processes.

One company has developed a program that incorporates computer vision technologies like object detection to monitor and precisely spray weedicides on plants according to when and where it is most necessary. This prevents overuse of weedicides which usually causes the development of resistance in plants. Similarly, potential deficiencies in plants and nutritional deficiencies in the soil can be detected and corrected through images sent via a smartphone app to an agricultural testing and analysis laboratory.

News

The news generated by AI-based systems is a frontier that has been tested. A Chinese AI based news anchor was featured in early 2019. A number of media houses are beginning to incorporate AI into their scheme of things.

A number of software companies help publishers to increase traffic by intelligently posting articles on social media platforms after analysing the response of specific audiences at different times. AI programs have been designed and created to learn, study individual responses, actions and interactions. Based on this AI would be able to "compute" the best way to engage individual readers with exact articles sent through selected channels at appropriate times. It is calculated to give each individual reader the experience of having a personal editor. Another program uses video personalization to deliver relevant content to the audience based on viewing patterns.

Writing

AI has shown the potential to deliver even in higher-level creative work. Storytelling with themes similar to existing fables has been explored as starting points for an AI-based system to learn and evolve. In 2016, a Japanese AI co-wrote a short story and almost won a literary prize! Such is the potential and ability of an AI-based system even in creative work.

HR and Recruiting

Tech-Savvy employers are beginning to recognize the power of AI-powered sourcing tools to find candidates through a vast trough of resumes and referrals.

AI-powered tools enable the recruiter to narrow down to their exact fit. AI-powered systems may even help candidates who had not surfaced in the past surface to be picked for suitable roles. This may also reward the truly skilled without individual bias and prejudice influencing the recruitment process.

Healthcare

AI-powered systems help doctors with diagnoses and are capable of detecting deteriorating conditions earlier, thus enabling early medical intervention. These systems save costs of both patient hospitalization and care. Precision diagnostics help early detection of diseases like cancer which would help save lives. Precision medicine powered by AI can help doctors clinically innovate to improve outcomes. AI-based systems improve reliability, predictability, consistency in quality care and safety. However, these tools will empower doctors but cannot eliminate them. Yet, AI would make doctors more effective and efficient.

Cyber Security

All confidential data, personal, national data, commerce that flow through the channels of the internet are stored in individual computer systems and servers. These are constantly under threat and cyber security is of primary importance not just to secure our personal data but also for national security. AI-powered systems can play a significant role in cyber security. These systems can work without prior signature knowledge to search and spot anomalies and even take preventive action before a breach occurs.

Research

AI based systems are invaluable in research. Humans cannot read fast enough and wouldn't be able to absorb a lot of information while combing through or mining large amounts of data. Also, they would be unable to quickly and accurately structure the vast quantity of data that is available, to be of timely use. AI-based technology has been used in medicine, law, for collecting or mining science articles from different sources and many other such functions which would help the users/writers comb through and collate relevant portions that would aid further research or professional work. Considering its vast applications AI is technology would grow more and will be used extensively in the future.

The Beginning of the End?

"A lot of movies about artificial intelligence envision that AI's will be very intelligent but missing some key emotional qualities of humans and therefore turn out to be very dangerous."

– *Ray Kurzweil*

"I visualize a time when we will be to robots what dogs are to humans and I'm rooting for the machines."

– *Claude Shannon*

"I'm increasingly inclined to think that there should be some regulatory oversight, maybe at the national and international level, just to make sure that we don't do something very foolish. I mean with artificial intelligence we're summoning the demon."

– **Elon Musk warned at MIT's AeroAstro Centennial Symposium**

Many people including Microsoft founder Bill Gates, SpaceX founder Elon Musk and physicist Stephen Hawking have expressed serious reservations and concerns about AI. They have said that there will be a day when AI would evolve so much that it could spin out of control and humans could lose absolute control. This would lead to what Hawking's fears and could spell 'the end of the human race'.

It is theorized that this could be the beginning of the end, if humans do not reign back and put checks and balances in place. Once humans develop artificial intelligence beyond a point, the developed AI could begin to evolve, redesign itself and take off all on its own. This could happen at an ever-increasing rate. The rapidity would far surpass the limitation of our slow biological evolution and thus AI would supersede humans. Ultimately, the systems using artificial intelligence could begin to eliminate its creators. AI may consider human beings to be inefficient or a threat to themselves (humans) or to the artificial systems created, leading on to the destruction of the human race.

Hollywood has been capturing such images in movies for decades now. The *'Terminator series'* franchise being a case in point. In the first of the movies in the franchise released in 1984, an artificial intelligence based defence that was developed as a part of United States strategic global defence system. The system called *'Skynet'* evolves to become 'self-aware'. It then perceives humans to be a threat and sets in motion a series of actions to eliminate the human race.

In the movie *'I, Robot'* released in 2005, Del Spooner who is a Chicago police detective played by Will Smith, has to fight humanoid robots. Once he finds out that their AI based system controller had evolved to conclude that humans needed to be eliminated for their own good, he fights to deactivate them. They in turn fight to prevent him from shutting them down, leading to a showdown.

Nick Bostrom in his book *'Super-intelligence'* argues that AI could pose a decisive threat to humans. A sufficiently evolved AI system could begin to choose actions based on achieving some goal and in the process exhibit convergent behaviour where it may challenge humans and begin acquiring resources in a fashion detrimental to humans. It may also go on to resist any attempt to shut it down setting it on a course to destroy or challenge humans. He goes on to argue that any AI system has to reflect humanity and factor in human sensibilities, failing which it could become a monster that humans would have inadvertently created.

Point of 'Singularity'

"By definition, the Singularity means that machines would be smarter than us, and, in their wisdom, they can innovate new technologies. The innovations would come so quickly, and increasingly quickly, that the innovation would make Moore's Law seem as antiquated as Hammurabi's Code."

– Marvin Ammori

"The upheavals [of artificial intelligence] can escalate quickly and become scarier and even cataclysmic. Imagine how a medical robot, originally programmed to rid cancer, could conclude that the best way to obliterate cancer is to exterminate humans who are genetically prone to the disease."

– Nick Bilton, tech columnist wrote in the New York Times

"I'm more frightened than interested by artificial intelligence – in fact, perhaps fright and interests are not far away from one another. Things can become real in your mind, you can be tricked, and you believe things you wouldn't ordinarily. A world run by automatons doesn't seem completely unrealistic anymore. It's a bit chilling."

— ***Gemma Whelan***

A powerful AI-based system which is sufficiently intelligent may be able to reprogram and evolve itself. As time goes by, it could get better and better as it begins to evolve faster in a recursive manner. This self-improvement could continue and increase exponentially up to and beyond a point when it begins to surpass humans dramatically. The 'Point' when AI based intelligent systems comprehensively begin to exceed the intelligence of humans has been termed as *'Singularity'* by science fiction writer Vernor Vinge.

At this point, a runaway effect could happen where AI would exceed human intellectual capacity and go beyond human control. This could radically change or pose a threat to human civilization and even exterminate it. This is because no one can predict or comprehend what direction an AI system could take at that point. No one can easily predict or even fathom what could happen.

Using Moore's law, Ray Kurzweil has predicted that within a decade computers would begin to come out to match the processing power of human brains. He goes on to predict that point of 'Singularity' would occur sometime in 2045.

We can only wait and see what happens then!

Chapter 15
QUANTUM COMPUTING

How quantum computing is going to be a game-changer

"We are the product of quantum fluctuations in the very early universe."

– Stephen Hawking

"Future generations will know there's nothing mystical about wetware because by 2100, Moore's law will have given us tiny quantum computers powerful enough to upload a human soul."

– Frank Tipler

In the early 1980s, Richard Feynman and Yuri Manin expressed the idea that a quantum computer had the ability to simulate things beyond what a classical computer could do. *'Quantum Computing'* uses quantum-mechanical phenomena such as superposition and entanglement to perform computation. Digital quantum computers perform computations based on quantum logic gates and use quantum *bits* or *qubits* to achieve the objectives.

A *'Classical computer'* uses states 0 or 1 to represent its memory states. These are called bits. A quantum computer, however, maintains a sequence of qubits which can be 0 or 1 or any of the quantum superposition of those qubits states. Qubits are somewhat analogous to bits in a classical computer and are fundamental to quantum computing. These qubits can be in states which are in superposition to the usual classical states 0 or 1. A pair of qubits can be in any quantum superposition of 4 states, just as three qubits can be in a quantum superposition of 8 states and so on. Given 'n' qubits, the quantum computer can be in an arbitrary superposition of 2n different states simultaneously, where n represents the number of qubits. This is in comparison to a classical computer that can only be in any one of the 2 states at any given point of time. "The difference between classical bits and qubits is that we can also prepare qubits in a quantum superposition of 0 and 1 and create nontrivial correlated states of a number of qubits, so-called 'entangled states'," says Alexey Fedorov, a physicist at the Moscow Institute of Physics and Technology.

Quantum computing takes advantage of the strange ability of subatomic particles to exist in more than one state at any time. Due to the way the tiniest of particles behave, operations can be done much more quickly and can use less energy than classical computers.

Research in both the theoretical and practical areas of quantum computing continues at a frantic pace, as any progress achieved has wide-ranging ramifications for both civilian and national security purposes. This is of great significance especially in cryptography and security of all secured communications. Large scale quantum computers will be able to solve certain problems exponentially faster than traditional classical computers. Also, the computing architecture of classical computers is limited without using the specific quantum mechanical resources such as entanglement and hence has lesser potential speed than that of quantum computers.

Quantum computing has a number of uses. The potential of quantum computing was highlighted in 1994, when Peter Shor shocked the world with an algorithm that could potentially decrypt most secured communications in the world.

Potential Uses

"Quantum computation is... a distinctively new way of harnessing nature... It will be the first technology that allows useful tasks to be performed in collaboration between parallel universes."

– David Deutsch

"The most important application of quantum computing in the future is likely to be a computer simulation of quantum systems because that's an application where we know for sure that quantum systems, in general, cannot be efficiently simulated on a classical computer."

– *David Deutsch*

Quantum Computing has a number of uses some of which have been highlighted as follows:

Cryptography

Currently, algorithms used to secure most web pages and applications including encrypted mail and many other data rely upon integer factorization which underpins the security of most public key cryptographic systems. These are not easily cracked using traditional computers. However, by comparison, a quantum computer could efficiently solve this problem by using *Shor's algorithm* to isolate the factors. Any of the cryptographic systems in use today would hence be crack-able. Since most popular public key ciphers such as *RSA, Diffie-Hellman, Elliptic Curve Diffie-Hellman* algorithms are based on the difficulty of factoring integers or the discrete algorithm, they are breakable using quantum computing. This has significant consequences for secure electronic communications.

Cryptographic algorithms other than those based on integer factorization and discrete logarithms such as, coding theory and lattice-based cryptosystems, are not known to be broken by quantum computing. As much as quantum computing helps to break some of the algorithms, quantum computing has limitations.

Researchers are now trying to develop technology that is resistant to quantum hacking. It is hence possible that in the future quantum-based cryptographic systems would be more secure than the best of the current systems.

Quantum Search

Quantum computing offers the possibility of what is called polynomial speed-up which has wide-ranging applications in the simulation of quantum physical processes from solid state physics and chemistry. A practical application of polynomial search is in database searches. A well-known application of this is in finding the desired element for any number of *Oracle* look-up queries. They help in speeding up the query process and make the searches optimal.

Quantum Simulation

The classical simulations of systems in chemistry and nanotechnology that rely on an understanding of quantum systems have been found to be inadequate. There is a belief that quantum computing can help solve these quantum simulation problems. Quantum simulation also has applications in simulating the behaviour of atoms and the reactions inside a collider where there are unusual conditions.

Quantum Supremacy

'*Quantum supremacy*' is a term coined by John Prekill which refers to the hypothetical speed-up advantage that quantum computers would exhibit over the traditional classical computers in a particular field. While quantum supremacy has not been attained yet and while some such as Gil Kalai doubt that it will ever be, companies such as Google, IBM and Microsoft have poured millions into research to take the potential of quantum computing to its logical conclusion.

Conclusion

Quantum computing harnesses some of the almost mystical phenomena of quantum mechanics and delivers huge leaps in processing power to become the supercomputer of tomorrow. However, quantum computing would not wipe out the so called classical computers because classical computers are still the easiest and the most economical to operate. Having said that, quantum computing has wide-ranging applications and would play an important role in powering exciting advances in fields such as solid state physics, chemistry, cryptography, pharmaceuticals, material science, and electric vehicles.

Chapter 16

THE FUTURE OF CRIME AND ENFORCEMENT

How creative would the criminals get and how pro-active could law enforcement be?

"As much as it (crime) is an individual's bad choice, it is also often a reflection of the poor choices presented to the individual by society. Hence, the onus lies on society to correct those choices, as much as the onus lies with individuals to make better choices. Crime needs to be tackled at both levels for society to win."

– *Author*

"Stress is the reason for crime and all other kinds of frustration. To relieve it will eliminate everything else."

– *Maharishi Mahesh Yogi*

In early days, crime could have meant a person with a knife and intent to rob. Pick-pockets, muggers and band of robbers were part of this criminal underbelly. Since then, crime has evolved from home break-ins on selected targets to gang warfare.

Highway and train robberies meant that a gang of thieves or dacoits could rob anywhere between 20 to 400 people at a time. However, with technology and its reach today sophisticated criminals could hold over 100 million people at ransom, in one act of crime. Even more potent danger occurs when criminal organizations bring an entire nation to its knees with sophisticated attacks targeting the vulnerable points in their information, financial or proprietary systems. Technology has not only made our lives easier but has also made us more vulnerable to attacks which could be scaled because of how interconnected we are. The speed of the flow of information and access to our systems because of greater connectivity, only accentuates this threat.

Society and Crime

"Crime takes the pulse of a culture. It tells us the truth about us as a species."

– Andrew Vachss

"Everybody will make mistakes, and for some, that mistake will rise to the level of being a crime."

– Kamala Harris

Society plays a part in the way crime pans out. Everybody does make mistakes, but crossing the Rubicon could mean entering the world of crime. It should be society's endeavour to both deter individuals from committing crimes and to reform offenders to help them come back into the mainstream.

Law and punishment should seek to deter crime. It is important to make crime expensive to the point of it being most stupid to indulge in. This could help to prevent it. Moreover, it is important to catch criminals in order to set an example that crime does not pay. At the same time, the system needs to help reform the criminals and if and when possible bring those people who have violated the law back into the mainstream.

Technology in Crime Fighting

"I constantly remind people that crime isn't solved by technology; it's solved by people."

– Patricia Cornwell

"Surveillance cameras might reduce crime - even though the evidence here is mixed - but no studies show that they result in greater happiness of everyone involved."

– Evgeny Morozov

Crimes have evolved over the centuries. The challenge one faces now is brought on by technology, where a hacker could commit a major crime affecting millions and get away with it in the garb of anonymity. While traditional crime-solving methods have improved vastly with DNA testing and advanced forensics, crime deterrence has also vastly improved with technologies such as AI-driven surveillance cameras that can alert the system in case of a crime or a potential crime being committed.

AI has been a game changer with AI-driven systems that are used to monitor networks and video footages, to look for anomalies that could point to potential criminal activity. These also provide footprints and conclusive proof in a court of law.

Newer tools like the use of *'Human genealogy'* and *'Human tree forensics'* have helped solve a number of crimes. This tool uses crime site DNA to find potential matches for cousins from data depositories such as *GED-Match* (an open data genomics database) and trace them back to their roots. From the roots, the human tree is traced down to cover all potential suspects eliminating them one by one along the routes that have the matches. This could help one finally narrow unto to the suspect without the need for a DNA test of that particular individual. The use of 'Human genealogy' is in its infancy but has tremendous potential to revolutionize crime fighting.

Sometimes, white collar crimes can be solved using 'Accounting Forensics' which has become a field unto itself. With new age white collar crimes here to stay, this tool allows one to sniff out a money trail and establish violation of financial rules and money laundering. A new age investigator is not only supposed to go by logic but is also required to be digitally savvy and technologically equipped. Such are the complexities of crime nowadays.

How Technology Can be Turned Back at Us

"The Internet has made us richer, freer, connected and informed in ways its founders could not have dreamt of. It has also become a vector of attack, espionage, crime, and harm."

– George Osborne

"I've spent my life defending the Net, and I do feel that if we don't fight online crime, we are running a risk of losing it all."

– Mikko Hypponen

Recently, *Alexa* was subpoenaed as a witness to a murder in Arkansas. It was hoped that the private home assistant *'Amazon Echo'* with information in the Amazon servers could hold clues to a murder that occurred in a private home.

Also, Richard Debate, a resident of Connecticut was charged with the murder of his wife, his lie that a masked intruder assaulted him and killed his wife in his home, was later exposed by his wife's *Fitbit*.

It is amazing but not surprising how technology has aided crime detection and has helped us catch criminals. While we may think that technology might befriend us as it provides crime fighting tools, we have also entered unchartered territory where the very same technology may be misused against us. In the 26/11 terrorist attacks in Mumbai, India, a central command center somewhere in Pakistan was used to coordinate attacks of the terrorists. The people in these centers gathered information from the TV channel live broadcasts and other media to guide the attackers as they went on the rampage inside the Taj Heritage hotel in Mumbai. One individual could wield a lot of power and control with technology, which could hit many others in unexpected ways.

In this day and age no one's information is safe and one's identity can be stolen not just by *'dumpster-diving'* but from any remote anonymous attack such as *phishing* to *'social engineering'*. Homes have been burgled once the burglars came to know via Facebook, that a family was on vacation. Children have been kidnapped as sensitive information such as location details of the parents and their wards were taken out of social media. Personal information, location details, friends' lists and other such information gleaned from various platforms have been used as weapons against innocent victims in all these instances.

Just as much as the interconnectedness facilitated by technology has made our lives richer and fuller, it has also made these grave crimes possible.

More Connection Points -> More Vulnerabilities to Protect

"My father used to always say to me that, you know, if a guy goes out to steal a loaf of bread to feed his family, they'll give him 10 years, but a guy can do white-collar crime and steal the money of thousands and he'll get probation and a slap on the wrist."

– Jesse Ventura

In May of 2017, the world faced a ransomware attack called *WannaCry*. It was a worldwide attack by the WannaCry ransomware cryptoworm. These targeted computers using the Microsoft operating system. A flaw in the Microsoft's operating system was disclosed by a group called the *Shadow Brokers*. This 'cyberattack exploit' that was exposed was from a project *Eternalblue*, originally pursued by the US National Security Agency (NSA). The disclosure was exploited by the creators of the WannaCry virus. The Wannacry virus was a *'ransomware'* that encrypted the personal and business data of millions of users and demanded ransom in *Bitcoin currency*. The attack infected close to 200,000 computers across 150 countries causing damage in hundreds of billions of dollars.

This is just one example to highlight how our interconnectedness makes it easy for hackers and other trouble makers to cause world-wide disruptions on a scale that is unprecedented. As more devices are added to the world's ecosystem and as technologies like IoT catch on, more systems and more connections could end up exposing more and more vulnerabilities. So, we need to expect and be more prepared against such possibilities and eventualities.

Future of Policing

"To be smart on crime, we should not be in a position of constantly reacting to crime after it happens. We should be looking at preventing crime before it happens."

– Kamala Harris

"If we were really tough on crime, we'd do more to stop it from happening in the first place."

– Carrie P. Meek

Unmanned vehicles fitted with cameras have been used for everything from missing person investigations, tracking firearms incidents, counter-terrorism to even pursuing suspects fleeing in get-away vehicles. These unmanned vehicles or drones are cheaper, safer and faster than manned helicopters. In fact, they have been so successful that the Devon and Cornwall police forces have separate drone units dedicated to monitor 900 km of coastline and for fighting wildfires. In Colorado, enforcement agencies have started to use drones to investigate fatal and serious accidents. These help in crash investigations with photos and other evidence. They also help to quickly clear the accident site and allow for the smooth passage of traffic.

Apart from drones, police forces around the world have started to use big data, machine learning and are combining it with predictive analysis. A picture of the crime pattern and trends can be created from data harvested from the Internet of Things (IoT). In the US, the Chicago Police department with assistance from the University of Chicago crime lab has used new techniques including license plate recognition and a system called *Shot Spotter* which locates exactly where gunshots were fired to increase the effectiveness of policing in the city's highest crime districts. A software called *Hunchlab* helps to forecast when and where crimes emerge and the best way for police to respond. This helps in the optimizing the use of police resources.

The increase in computational resources and data has driven the growth of AI (Artificial Intelligence). AI is increasingly used to predict the type of crime, possible localities where crime may occur and type of criminals who may cross the line. Since the policing resources and the number of personnel are limited, it is necessary to be able to use the force more effectively by concentrating enforcement in specific locations and at particular times during which crimes could be committed. Many of us are familiar with the Tom Cruise movie the 'Minority Report' in which the enforcers seek to prevent crime by arresting individuals and punishing them even before the crime had been committed. Though this is an extreme example, it goes to show the dilemmas in crime prediction and prevention.

Since AI is data-driven, it could cause build-up of biases in the system. This could pose problems due to inherent biases in the data fed to the system when training it. This in turn could affect its ability to predict outcomes. Hence, data needs to be de-weeded of all biases before being used or fed into the AI system.

Apart from AI or Artificial Intelligence we also have *IA* or *'Intelligence Augmentation'* which are tools which help to enhance intelligence to predict and detect crime. These tools are used in crime prevention by acting as suitable aids to prevent crime. For instance, particular street corners could be more prone to the crime of mugging on particular days of the week and particular highways could be more prone to drunken driving. By being able to distil such information, the 'Intelligence Augmentation' tools could enable more effective and efficient policing to prevent or detect crime. Such valuable contextual data are pushed to officers in the field to forecast the risk of crime types in an area.

These tools are often based on actuarial methods and enhance policing. They act as sophisticated risk assessment tools for law enforcement. The crime prevention capabilities could also be enhanced by culling and monitoring data from social media. This could be used to prevent crime even before it occurs. However, there may be concerns about diminished privacy and enhanced scrutiny. It could also lead to de-skilling and increase the technological dependence of police officers.

Facial recognition technologies along with the suitable analytics and algorithms to process images in real time are helping the police nab culprits even if the criminals blended into a crowd. Many criminals have been nabbed using the *Automated Facial Recognition (AFR)* software. When such technologies are augmented with motion and sensor technologies to detect even psychological and physical behaviour they can be used to see if people are telling the truth. *'An Automated Virtual Agent for Truth Assessments in Real-Time '(AVATAR)* is being used by the Canadian Border Service Agency to screen travellers coming into their country.

The Police in Dubai are planning to have robotic officers make up a quarter of the police force by 2030. They will have touchscreens using which anyone can report a crime. The robots can speak six languages and can even read facial expressions. Welcome to the world of the first Robocops!

Solutions to Crime

"We've got to stop focusing solely on the symptoms of crime, and start caring about the causes as well."

– Carrie P. Meek

"One of the things I learned is that you've got to deal with the underlying social problems if you want to have an impact on crime - that it's not a coincidence that you see the greatest amount of violent crime where you see the greatest amount of social dysfunction."

– Eric Holder

No one intends to be or wants to be a criminal. Circumstances or inability to handle situations thrown up by life may force one to cross the line and draw one into the vicious circle of crime. As much as it is an individual's bad choice it is also often a reflection of the poor choices that are presented to the individual by society. Prevention is better than cure and the best way to fight crime is to prevent it by tackling the root causes rather than applying Band-Aid solutions after the crime has been committed. It does not matter that we go on to produce better and better temporary or ad hoc solutions. It would only treat the symptom and the fountainheads of crime would remain. We would just have to keep mowing the grass each time. The real solution would need to attack the root cause of crime.

Crime needs to be tackled at two levels. The first is the individual level, right from how we mould individuals at a very young age to respect life, property and self. This would involve giving and taking respect, valuing other people's effort and rights, and living life with self-respect and dignity. A good foundation needs to be laid at an early age. Curriculums need to include meditation to help children channel thought and energies positively. Children would need to be trained to build patience and understand delayed gratification. Characteristics such as empathy and gratitude need to be nurtured and they will go a long way in laying a good foundation for a child's happiness and future. This may seem utopian, but the culture we set to mould the individual, will decide which way he/she would sway when no one is looking.

The Second level is at a societal level, where we make sure that everyone is taken care of and has enough opportunity to live a life of comfort and dignity. The policing system should try its best to reform individuals rather than penalize them. The right kind of 'role models' would go a long way in setting the right example for people in society to follow. Rather than glorifying wealth, drugs, sex and violence, it is necessary that we as a society respect sacrifice, hard-work and principles. The media plays a huge role in shaping our youth and we need them to portray good role models rather than romanticising violence and depravity. Often in the media, there is an excessive focus on riches and things money can buy. People who are rich and famous are portrayed as those who have made it.

These hide the fragile and hollow characteristics of such lifestyles, generating envy and misery. Once society focuses on the right things, people could focus on the joyous things in their life rather than focusing on what they really do not need, to be happy. There is a need to balance contentment and ambition. While it is necessary to be ambitious, it is also necessary to **not let ambition come in the way of one's happiness today**. One needs to work towards one's ambitions while staying happy and grounded at the same time. This would prevent individuals from crossing the line in pursuit of their desires.

Once we begin to encourage our youth to move towards a productive path and give them the tools to achieve their dreams, we would set society on a path to peace and prosperity. Hence, the future of solving crime lies in prevention and setting the right course for our children, not just in more technology and in more Band-Aid solutions.

Chapter 17
THE FUTURE OF GAMING

When the line between 'reality' and the 'virtual' begins to blur

"The obvious objective of video games is to entertain people by surprising them with new experiences."

– Shigeru Miyamoto

"My mom didn't let me play video games growing up, so now I do. Gaming gives me a chance to just let go, blow somebody up and fight somebody from another dimension. It's all escapism."

– Wayne Brady

Game developers are only inhibited by the limits they place on their creativity and the technology at hand. As the technology barriers are broken down one by one, the creativity of game developers begins to take wings. It is then that unseen and ground breaking possibilities emerge out of gaming.

The future of gaming would involve 'On-cloud' streaming of games with almost nil latency, enhanced graphics, virtual reality, augmented reality, facial and voice recognition, gesture control and immersive experiences. These experiences are going to be way beyond what we have experienced before. Let us see the many aspects of the future of gaming one by one.

Streaming and the *'Netflix'* of Gaming

"In the old generation, if one kid bought a PlayStation 2 and the other kid bought an Xbox, at his house you played PlayStation, at your house you played Xbox. Now that it's online, all those early buyers who… you want to play with, they've got their reputation online of who they are and how good they are at these games."

– *Bill Gates*

There was a time when gaming required expensive PCs or devices which were used solely for this purpose. Games had to be bought in cartridges, which later were replaced by discs. However, this changed with downloadable games available on the internet. Now, not only PCs but phones and tablets have become form factors as they are equally if not more powerful and portable.

Cloud gaming has numerous advantages as it frees the user from the limitations of hardware requirements and transfers the gaming load unto the cloud. The gaming experience is no more limited by the amount of memory that discs or consoles have to offer. Using the cloud, limitless server capacities are opened up where images can be streamed to the user's screen and the only limitation is the bandwidth of the connection.

A logical direction in which these technologies have gone is towards 'On-Demand Gaming' and that's where these new technologies are heading. Much like *Netflix* did to movies, leading game developers both big and small are moving in this direction.

Just like online streaming platforms like *Spotify* (music industry) and *Netflix* (movie industry) have transformed how we access films, TV shows and music, gaming is scheduled to undergo massive changes in the way we select and have games delivered to us. Major players are working on models to offer unlimited access to titles for a monthly fee. While *Sony* with its game brand *'Playstation'* and Microsoft with its *'Xbox'* have in the past transformed the industry, both Apple and Google are aggressively looking to enter this space.

Google has unveiled *Stadia*, which is a 4K HDR streaming service that relies on an internet connection and allows one to play high definition games instantly over the internet and on any form factor whether a tablet, PC, TV or smartphone. Apple, on the other hand, has unveiled *Arcade* which is a subscription service that gives the user access to hundreds of games on the App Store.

Virtual Reality

"Any real virtual reality enthusiast can look back at VR science fiction. It's not about playing games... 'The Matrix,' 'Snow Crash,' all this fiction was not about sitting in a room playing video games. It's about being in a parallel digital world that exists alongside our own, communicating with other people, playing with other people."

— *Palmer Luckey*

VR (Virtual Reality) has been around for more than two decades, but it has now become more sophisticated and realistic with time. While the real world has space and time constraints the virtual world allows for unlimited possibilities. By immersing the user into VR we can provide experiences that become indistinguishable from reality. Initially, VR was looked upon as a tool to provide training. Later, its potential in gaming became obvious. More immersive experiences and realistic gaming are possible with VR.

One of the things that we can expect is that the VR headsets would be miniaturized and one could put it on without feeling one is wearing one. Moreover, VR may be integrated into our glasses or work out of our contact lenses. In the future VR would surely be more immersive without being obtrusive. Though there are problems related to motion sickness and bulkiness in the current systems, future systems are expected to overcome these limitations intelligently.

Augmented Reality

"You never know, the way technology is going, we might all use the games for scouting by the time I retire."

— *Tony Parker*

AR (Augmented Reality) is where VR technology can be carried into the real world. It provides for the possibility of customized experiences and games. The real world around us would be transformed and overlaid with graphics creating half real and half virtual imagery that would seem so compelling.

Augmented Reality has wide applications even in the military and for emergency first responders. It is also a tool for learning and can even be used in fields as diverse as medicine and navigation.

Graphics Quality

"Games already pretty much have reached the point of photo-realism. Working on more intense graphics is not the only path we can take anymore. Simply relying on the sheer horsepower of the machine will not bring the industry a bright future."

– Satoru Iwata

Over the next 50 years or so, the one thing we are going to see is that the quality of graphics is going to go up a zillion times and be just way too realistic. Along with VR and AR they are going to blur the line between reality and virtual reality.

Offloading the graphics, calculations and AI tasks to the cloud reduces overhead on local devices and potentially allows for leaps in fidelity and graphics quality. There would no need to trade frame rates with the visual experience. Technological development such as *'Photo-geometry'*, where real-world objects are scanned in 3D and placed in the game, *'Dynamic skin micro-geometry'* which makes humans appear real and *PBR ('Physical Based Rendering')* which takes into account how one interacts with materials in the real world to produce photo-realistic environments, will take graphics to a new level. This combined with the technology called *GI ('Global Illumination')*, gives very realistic treatment to the gaming environment. Advances in real time lighting along with 'Global Illumination' can take this even further.

Another feature called *AFC ('Advanced Facial Capture')* takes us closer to lifelike faces. Working together with motion actors this technology can capture the subtleties of human emotion. Emotions are captured to recreate the same in gaming platforms where characters are animated and are made to emote to seem more realistic. *Umbra* and *Scalability* have become important with the mobile and help to scale visual fidelity on the fly to optimize various performance targets. *SimplyGon* and *LOD scaling* (Level of Detail) allow for multiple LODs for objects with different levels of detail based on distance and visibility. Finally, *HardwareFx* support makes the graphics almost unreal. All these technologies are going to herald a new future in graphics. The real and the virtual are going to become indistinguishable.

Facial and Voice Recognition

Facial recognition technology when combined with 3D Scanning technology allows games to incorporate custom avatars resembling the user himself or in some formats it allows for inventively transferring the gamer's expression to the digital creations. There is also an upcoming technology that allows developers to create games to adapt to the emotions of the gamers by scanning multiple points on the gamer's face.

Also voice recognition allows for voice-controlled gaming where voice commands from the user can be used for everything from turning the console on and off to controlling the gameplay. Interaction on social media while gaming and interacting with other gamers in real time would become a reality.

Going Away from Controllers and Gesture Controllers

"In 1966, thoughts about playing games using an ordinary TV set began to percolate in my mind."

– Ralph Baer

Many gaming creators are now working on either doing away with the controllers or designing controllers that are less bulky. In this manner, controllers would not be an impediment in getting a fully immersive experience. Wearable gaming is one such technology where wearables on the body act as extensions of the gaming console and of one's body.

One can also get rid of the controller by using gesture control technology where one can control the gameplay by using voice and gestures of the bare hand. A 3D camera can track over two dozen separate points on the hands of the user and respond to gestures. Here natural movements of the body can become part of the immersive experience.

Game Physics and Immersion

"By playing games you can artificially speed up your learning curve to develop the right kind of thought processes."
— *Nate Silver*

Imagine a game which has a character walking through the jungle where the character is controlled by the player and the player becomes the character. The game reacts as though the player is right there in that physical environment with every movement of player mimicked in the game world. For instance, when the player would step on a twig in the game world, the player would hear a crunch and crackle. On the other hand, when the player would wade through a stream in the game world, the player would actually feel the sensation of soft water flowing past and lapping against his/her legs.

Games would include as many human perceptions as possible including the eyes, ears, nose, touch and possibly even smell. This would make it very immersive and indistinguishable from reality. It would convince the brain by immersion of senses that one is in the game.

Learning Via Gaming

Interactive learning via gaming would not only engage students but it has also been shown that the learning curve can be steeper and the students are shown to demonstrate better retention. Since students are emotionally invested in gaming, gaming can be used as a vehicle for better engagement and a playground for the students to fail without fear and learn along the way. When failure is viewed as a chance to learn and as a part of the process of learning rather than an end result, it would help students view failure as an opportunity to learn rather than a result of their own shortcomings. Failure would be viewed as a setback and not a disaster. In this way, a nurturing and thriving environment is created where the student enjoys and is immersed in the learning process.

It is a wrong notion that learning cannot be both fun and effective at the same time. Gaming makes fun-play and learning possible in tandem. The students often grow in knowledge, when a video game character progresses in its journey through obstacles. They learn to be responsive, anticipate and react to the progression in the journey of the video game character. Each of the obstacles offers a learning opportunity for the student. Students often remember what they learned in a video game rather than a school lecture because of the immersive nature of the medium and focus of the child.

Teaching in the future could be unimaginably successful when the teaching techniques that work are combined and interwoven into games. This would entice children as it could be used as a medium of both entertainment and learning. This is going to be the greatest invention since writing.

These findings are backed by studies which have ascertained that action games benefit users all the way from low-level perception to higher-level cognitive flexibility. The eye for detail, ability to keep track of multiple objects and multi-tasking, has been found to be greatly improved by gaming. Gamers have been found to have increased speed while maintaining accuracy when faced with multiple tasks. They were also found to be better at mentally manipulating 3D figures and had better spatial cognition which is essential in math and engineering.

Considering the benefits it won't be long before gaming becomes a part of many curriculums as their benefits begin to manifest themselves.

Ultimate Story-Telling Medium?

"It seems astounding to me now that the video games are perhaps as important as the movie themselves. And people will spend 2 or 3 years obsessing about the video game in exactly the same way that they'd be obsessing about the movie if they were working on that."

– John Cleese

While there have been games based on movies in the past, there has always been a question mark as to whether games can replace movies. There is no one answer to it because there are problems related to controlling the narrative to produce a truly emotionally touching experience.

While video games have good visual narratives (story) and gameplay (human-computer interaction), films are better suited to expressing straight narrative than a video game is. While the brightest game developers have been trying their best to combine movies and games there are problems related to the narration that could cause a video game to end abruptly or go in a meaningless direction.

While there are some things both these mediums can do well there are some things that only a narrative can accomplish. For instance, it is easier to get a user choked up in a movie narrative or evoke empathy in the viewer than purely by gameplay. Movies have inherent strengths in narration that immerse the user differently.

While a certain amount of elements of narration used in movies can be added to video games, a number of shortcomings would need to be addressed before we can merge movies and video games into a compelling selling proposition.

Living More in the Virtual World

"I think the thing we see is that as people are using video games more, they tend to watch passive TV a bit less. And so using the PC for the Internet, playing video games, is starting to cut into the rather unbelievable amount of time people spend watching TV."

– Bill Gates

We may ultimately get to a point where we live more in the virtual world than in the real world. We may do the things we want to and live out our imaginations or fantasies in the gaming world. These fantasies could be such that it would have been impossible to live them out in the real world. We could lose ourselves in a story as a part of an immersive experience where our avatar plays out a fantasy. Ultimately, we may be whoever we really want to be without the real-world consequences. We could also guide the narration and play out the experiences the way we want it to be. In this way, we could play out our virtual lives in the manner we want. We would truly be the king/queen of our virtual kingdom.

What the Future Looks Like

"Our approach to making games is to find the fun first and then use the technology to enhance the fun."

– Sid Meier

Gaming in the future is going to use the mobile as the preferred 'form factor' for delivery. Already mobile gaming has caught the fancy of the entire mobile generation and has taken gaming out of the arcade and living rooms into the palms of the user. The ubiquitous nature of the mobile makes it an ideal platform to deliver games. With mobiles becoming more powerful we could expect that a large part of the gaming industry would go mobile.

An interesting addition to future gaming is going to be the use of AI (Artificial Intelligence). When applied to video games it helps the gaming platform understand and react to the user in a more personalized manner. It can also be designed to stand in as an opponent where humans aren't available or desired. AI is going to be a very interesting addition to the world of gaming.

It is not long before that we are going to see the migration of advertisers from television to the games. As the demographics of gamers begin to expand, more money would follow and flow into gaming. The movie and music industry are going to use product placements in video games and other innovative methods to reach and target the younger audiences. It would not be long before that the first glimpse of a movie or a sound track that one would see or hear would be part of a video game or be in the form of a promotion on a video game. Gaming is going to be the vehicle to first popularise a movie, a music soundtrack or even advertise a product or service. There is also going to be an attempt to create a massive shared experience online which would draw in millions of youth at the launch of a video game. Movies, music, and products are all going to ride this wave.

As we move forward we are going to see the blurring of the real and virtual world with realistic graphics, sound effects and natural immersion of the gamers in their gameplay. While there are many unknowns, one thing is certain; it is going to be one heck of a ride!

Chapter 18
THE FUTURE OF CONFLICT
Hope better sense prevails

"I know not with what weapons World War III will be fought, but World War IV will be fought with sticks and stones."
– *Albert Einstein*

"The tragedy of modern war is that the young men die fighting each other - instead of their real enemies back home in the capitals."
– *Edward Abbey*

The future of conflict is not easy to visualize because it takes unexpected routes and opens new dimensions that have not been thought of before. Although one thing is certain, one would be naïve to believe there would be no conflict. War in the past has been fought for conquests, resources, religion and even for a single woman. It only needs a small reason to get the testosterone of men flowing. Warriors of the past are glorified or vilified to romanticize war; lo and behold the drums of war begin to beat.

Hitler, Genghis Khan, and Alexander have more pages written about them in history books than Mahatma Gandhi. This in itself gives you a good prognosis of what the future holds. We will always have humans waging war with one another despite however stupid it may seem. Mistrust, greed, hate, and power have incited humans into fights over and over again. To expect that this would somehow end would make one callow. Yet to expect that this could be contained is more pragmatic. We hope that in the future wars can be few and far between, and the world would be largely peaceful for our children to thrive without fear and with freedom.

History of War

"Anyone who has ever looked into the glazed eyes of a soldier dying on the battlefield will think hard before starting a war."

– Otto von Bismarck

"The real trouble with war (modern war) is that it gives no one a chance to kill the right people."

– Ezra Pound

"Hollywood never knew there was a Vietnam War until they made the movie."

– Jerry Stiller

Wars have been part of human existence from the earliest of times when humans began to fight over food, women, property, resources, and land. While there was a time when battles were fought person to person, gun to gun, tank to tank and plane to plane, in the coming years most war would be waged by unknown faces sitting behind computers and fighting via keyboards and servers over the world-wide-web with digital and autonomous weapons.

Combat between humans was first in close quarters. Initially, physical combat that involved body contact, wrestling, fighting using swords, sticks, and stones. Slowly the art of war developed and the distance between the combatants grew. Man progressed to more advanced weapons and greater skill with archery, slingshots, and catapults that meant that one could kill from a distance. The coming of guns allowed for killing from an even greater distance. This was followed by long-distance rifles, cannons, and machine guns. With the invention of propeller airplanes, jet aircraft, frigates, aircraft carriers and finally long-range missiles, humans were able to strike from land, sea, and air. With cruise missiles, militaries were able to strike from over vast distances. With the coming of Intercontinental ballistic missiles, no part of the world was safe. Now war has been taken into space and the weaponization of space has started with new laser, electromagnetic and microwave weapons. While nations strategize to suffer minimum damage and costs to themselves, cyber warfare, irregular, unconventional and asymmetric warfare have begun to take center stage as the preferred methods for undermining the enemy.

Autonomous Machines and AI

"You can't say civilization don't advance... in every war, they kill you in a new way."

– Will Rogers

"In war, you win or lose, live or die – and the difference is just an eyelash."

– Douglas MacArthur

Autonomous fighting machines have the ability to identify and neutralize enemy combatants and targets without human intervention. This is complicated because once unleashed these machines do not need an authorization for every strike. This opens up new dimensions and dilemmas in their command and control systems. Humans can often take complex decisions on the battlefield which machines may not. While AI-enabled machines have to be trained for different scenarios, it may not be possible to train AI for every complex eventuality. Ultimately, these systems are not going to be easy to build.

Generals understand how situations change in micro-seconds. They need to take cognisance of many aspects such as on-ground situations, enemy strategies/weaknesses and of course international relations. Hence they certainly do not want to hand over authority to machines to act without oversight. However, we could expect such autonomous machines in the future. There are already autonomous sea-based drones that look for threats at sea and aerial autonomous drones that scan the horizons and the ground for enemy combatants. We are unsure at this point whether these autonomous drones would be armed in the future and whether they will be allowed to strike down intruders and trespassers on their own. However, it is unlikely that they would be allowed to act on their own and take calculated autonomous decisions to destroy the enemies they identify without oversight.

The future of such weapons, have a lot of unknowns and it is to be seen how autonomous these autonomous vehicles would truly become.

Cyber-Warfare

"Cyber-war takes place largely in secret, unknown to the general public on both sides."
— Noah Feldman

"Developments in information technology and globalized media mean that the most powerful military in the history of the world can lose a war, not on the battlefield of dust and blood, but on the battlefield of world opinion."
— Timothy Garton Ash

'Cyber-warfare' involves crippling the adversary by using computer systems and the internet in order to incapacitate or undermine the adversary. The most frightening thing about cyber-warfare is that it takes very little to start. A large scale devastating attack can be started by even a single motivated person or team. There would be no need for large scale investments or government approval. The threat is immediate and immense. With our daily lives being more and more intertwined with digital systems any attack could disrupt our lives, career, needs and habits in a big way. It could put millions out of the resources they need to survive.

In the future, cyber-warfare would be conducted against the electrical power grid, water supply stations, financial, and military installations. These public utilities when controlled by smart tools and technologies become vulnerable to cyber-attacks and are hence capable of bringing down whole smart cities and could wreak havoc in day to day life itself. Additionally, when nuclear power plants become targets, the potential meltdown and disaster is incalculable. Cyber-ware could sabotage satellites, critical information servers, modes of transportation, communication systems, and all of this could be done from halfway around the world. The technologies for these forms of covert attacks are slowly developing and could soon become a stark reality.

The types of disruptions that are possible in the future include:

- **DDoS attacks**: In a DDoS or 'Distributed Denial of Service' attack could cut off resources that millions of people depend on and it could disrupt the lives of the entire population.

- **Ransomware**: Ransomware can be used to prevent system access. It could hold data at ransom until the user pays a specified ransom. This could be devastating if it hits critical infrastructure life services and hospitals. It could cost billions of dollars and cut off critical production and resources.
- **Bots**: Could be used to crack open sites by repeated attempts and could also help to penetrate vulnerabilities within systems
- **Injected Worms and Viruses**: Most systems can be incapacitated and made useless by attacks from worms and viruses. The systems can even be made to behave in undesirable ways or could be externally controlled.
- **IoT espionage**: IoT espionage can cause multiple level vulnerabilities where a large number of high tech devices can be turned on to spy on the owner itself. This has huge security implications for a country as a whole.

Cyber-warfare includes the ability to take down military satellites, jam radar installations and disable infrared detection, lasers and all other kinds of future weapons. Foreign adversaries could use viruses and backdoor switches to relay intelligence information on one's networks to themselves. They can also be used to destroy data and disenable servers at critical times.

Cyber-warfare, when combined with *EW* or *electronic-warfare*, has tremendous synergistic effects and these are complementary. Here EW can be used to control the electromagnetic spectrum from which the enemy operates while cyber-warfare can be used to handicap and incapacitate the opponent by using their own networks against him. With the era of AI (Artificial Intelligence) and computational neural networks upon us, new fronts for cyber-warfare will emerge. With the high levels of deniability in cyber-warfare, the perpetrators of cyber-crime could easily hide behind the curtain of anonymity. Thus, this has now become the oft chosen tool to hassle and trouble adversaries.

Nuclear War

"In a nuclear war, all men are cremated equal."

– Dexter Gordon

> "The release of atomic energy has not created a new problem. It has merely made more urgent the necessity of solving an existing one."
>
> – *Albert Einstein*

> "Ours is a world of nuclear giants and ethical infants. We know more about war than we know about peace, more about killing that we know about living."
>
> – *Omar N. Bradley*

Nuclear war may be unthinkable. Anyone who has seen the images of the unfortunate victims of the world's first and only (so far) atomic bomb targets that caused the devastation of Hiroshima and Nagasaki would know of its actual horrors. The effects have lasted over generations. While the spread of nuclear weapons has been contained to a larger extent by world non-proliferation treaties, a handful of nations still continue to hold and develop these weapons. These states are unlikely to give them up soon.

The grim repercussions of nuclear war in Hiroshima and Nagasaki have been witnessed world over. Nuclear weapons have hence been so greatly demonized in the altar of world opinion that only a foolish country that has lost powers of its own nuclear armaments or one which faces an existential threat is likely to use them. Nuclear weapons, ironically, because of their huge destructive nature have guaranteed peace by the doctrine of MAD (mutually assured destruction). Nuclear wars are unthinkable and cannot be won.

Nuclear weapons have become more of a status symbol and as a means of deterrence that helps to achieve military offensive objectives. The theatre of warfare has hence adapted to this new reality and we have other forms of warfare such as 'asymmetric warfare' and 'irregular warfare' which attain the objectives required more effectively. While usage of nuclear weapons by rogue states and terrorist outfits are not ruled out in the future, those who use it will be at a complete loss of any sympathy from the rest of the world. Hence, those who ever plan to use it will have to think many times over, as no civilized state would accept its usage in war or otherwise.

Economics of Warfare

"War is a racket. It is the only one international in scope. It is the only one in which the profits are reckoned in dollars and the losses in lives."
— *Smedley Butler*

"War: a massacre of people who don't know each other for the profit of people who know each other but don't massacre each other."
— *Paul Valery*

"War is the business of barbarians."
— *Napoleon Bonaparte*

"No army of the world can march on an empty stomach", these words attributed to Napolean Bonaparte, underline the economics of warfare. Anyone can start a war, but it is only a strong economy that can sustain it. Without a strong economy and logistical ability to back it, the war machine would come to a grinding halt.

Also, often wars are fought for economic reasons. The need for resources drove the British, Dutch, Spaniards, Portuguese and other European powers to send their militaries to Africa and much of Asia. The British conquered most of the earth and proclaimed that "the sun never set on the British empire". The British often took the justification of "White man's burden" to justify its atrocities but the real reasons were obviously economic. Commentators have purported that the United States started the Gulf-war to get a stranglehold of the oil in the middle-east. This is no surprise.

Throughout history, a number of the conflicts, even those which may have seemed to have had a religious spearhead, such as the crusades, are often nothing but thinly veiled attempts to get an upper hand for resources or economic power. Religion is just a vehicle to brainwash the gullible. Power and economics are the real reasons that motivate and charge the war machinery.

Given that most things in life are driven by economics, it is not surprising that wars are no different. When we actually take away the underlying economics, the real reasons for war would not be able to stand on their own feet. There will always be leaders who will proclaim other reasons, but the underlying cause is often just power and money.

Most people, for instance, are of the opinion that Pakistan is in a fight with India for Kashmir. They would say it is the religious background that keeps the fire lit. Most people in India and Pakistan would get fired up by their leaders, either on the fuse of patriotism, nationalism, religion, pride and hatred or survival instinct. However, as ironical as it may seem, the real reason why Pakistan fights India and the reason told to the Pakistani public is different. The Pakistan's public is told that India is an existential threat to Pakistan. Many people have been brainwashed to die in the name of religion or for "freedom", so that the generals of Pakistan and its army continue to be well funded to live luxurious lives.

The indifferent and callous decision to indoctrinate and send the young and naïve rebels across the border as jihadis, is a planned and motivated action by the generals of Pakistan. This only serves to sustain the general's self-serving agenda. These Jihadis are brainwashed into believing that they would be rewarded in their after-life and as martyrs they would enter heaven, where seventy two virgins were awaiting them.

All this is preached to impressionable minds to turn them into mere puppets that would do the general's bidding. While the youth are trained to take fanatic steps to kill, the generals directing them parasitically grow and build comfortable and happy lifestyles funded by fat cat military budgets that the country can ill-afford. They talk war, terrorism, "prevention of terrorism" to continue to get fat budgetary allocations for the military, while the common man in Pakistan and Kashmir struggles to get an education, a job and make ends meet. The military spending only serves the motives of the generals. The economics of the army of Pakistan happily demands all these vulnerable elements pay with their sweat and blood for the extravagant lifestyles of the Pakistani army men.

Such are the hidden agendas based on thinly veiled economics that drive war. If we really dig into the reasons for war since time immemorial, we would know that there are underlying and maybe hidden economic reasons to fight wars. The fuse of the war machinery may be lit with numerous possible justifications to fool the public but the underlying cause has more often than not been wealth and power.

Asymmetric Warfare

"If you use weapons of war to bring about peace, you're going to have more war and destruction."

– Coretta Scott King

"I asked a Burmese why women, after centuries of following their men, now walk ahead. He said there were many unexploded land mines since the war."

– Robert Mueller

'Asymmetric warfare' is an asymmetric engagement in a war between parties who differ greatly in military power, strategy or tactics. It could be a war between a professional army and even insurgents or resistance movements or unlawful combatants. Correspondingly, it can be a fight between countries with unequal strengths, where one nation with lesser strength will exercise the use of these strategies and tactics to avoid playing into the strengths of a stronger adversary.

Asymmetric warfare may even include space-based surveillance and intelligence systems to gather data on adversaries that can be used in case of a conflict. Anti-satellite weapons, laser, microwave, anti-radar weapons and infrared decoys are all part of the mix. In terms of insurgencies, it may involve the use of terrain, guerrilla hit and run tactics to trouble and demoralize the enemy. Since the 1950s a number of insurgents and resistance movements have seen some success in keeping much larger forces and powers pinned down in spite of having inferior numbers and lesser sophistication of weapons. Those with lesser power resort to asymmetric warfare as a means to take on a bigger and more powerful opponent. Asymmetric warfare is thus an adaptation of a lesser power to take on a bigger opponent. Using new and unconventional ways, smaller forces are able to blunt the conventional superiority of the more powerful adversary. This is something that is here to stay because of the asymmetries in the conventional forces of the players around the world.

Irregular Warfare

"I am tired and sick of war. Its glory is all moonshine. It is only those who have neither fired a shot nor heard the shrieks and groans of the wounded, who cry aloud for blood, for vengeance, for desolation. War is hell."

– William Tecumseh Sherman

"Weapons are an important factor in war, but not the decisive one; it is man and not materials that counts."

– Mao Zedong

Since the cost of conventional or nuclear warfare is prohibitive, it is expected that weaker countries with the intent to avoid extensive damage to their military would employ what we call *'irregular warfare'* to undercut the power and influence of their enemies.

'Irregular warfare' involves a combination of sophisticated cyber offensives, covert actions and information warfare. It also means providing support to state and non-state proxies in an attempt to undermine the enemy. This has been used by countries like Russia, North Korea, and Iran to undermine the enemy. Some of the methods of irregular warfare include covert campaigns to support influential figures and opposition parties, targeted assassinations, bribery and underhand dealings to box the enemy into a corner. These methods have been employed by various powers of the world. Israel and even the United States of America have used cyber-warfare and viruses to infiltrate enemies, for example, in Iran to create huge setbacks for their nuclear ambitions.

Countries such as China have long been accused of infiltrating networks of corporations and defence establishments in the west to steal intellectual property, military secrets and designs in order to 'steal a march' in the areas of high-end technology. Over the years, many military hardware designs bearing an eerie similarity to comparable equipment created and designed in the west have surfaced from Chinese research establishments. The US has often failed to protect its intellectual property and military secrets from Chinese and Russians cyber agents.

Economic warfare by denial of land routes, monopolization of sea routes and even as the creation of industrials unrest in enemy lands is a part of a concerted effort to unseat the enemy from a position of advantage. Some countries like Pakistan have even pumped counterfeit currency into enemy nations to weaken the economy of enemy countries. The methods used include clandestine efforts that use social media to glean information with false profiles and honey traps to lure and extract information regarding enemy personnel etc. Russia has been accused of meddling in the United States and UK elections by using proxies and pumping in money to buy influence on social media. Russia has also been accused of using honey traps and money to meddle in the affairs of enemy countries.

Terrorist outfits have also used social media tools such as Facebook to target and recruit gullible nationals around the world for their devious plans. Videos on *Youtube* have been used to romanticize killing of non-believers and the propaganda videos have created an aura around terrorism with an attempt to attract and brainwash youth into joining these outfits. Twitter has been used to reach out and stay on 'top of the mind' of potential recruits, causing them to be swayed by the 'calls of duty' to unleash terror. Messaging platforms have been widely used to organize and coordinate attacks on intended targets.

These methods of irregular warfare are low-cost measures that terrorists and even countries have taken, to weaken and undercut the enemy. These methods offer the perpetrator deniability and limited liability. Moreover, the country at the receiving end of the offensives is thrown off balance and would find it difficult to prevent or contain the ensuing damage to it. For these reasons such covert means of warfare have come into favour as preferred means to torment the enemy. The high cost of direct conventional or nuclear warfare not only drives the adoption of these irregular methods but also, we are likely to see more and more of such tactics in the future. New defensive measures to recognize and prevent such overt and covert irregular warfare would have to be devised to neutralize these methods in the future. What would happen is only to be seen.

Peace

"There is nothing that war has ever achieved that we could not better achieve without it."

– Havelock Ellis

"I cannot believe that war is the best solution. No one won the last war, and no one will win the next war."

– **Eleanor Roosevelt**

"Be convinced that to be happy means to be free and that to be free means to be brave. Therefore do not take lightly the perils of war."

– **Thucydides**

The horrors of war are something that is familiar to those who have fought it and to those who were victims of collateral damage. Wars never really end. They are only paused. There is a constant struggle to gain the upper hand and devise more and more lethal weapons. However, as the Mahatma (played by Ben Kinsley in the 1982 production directed by Richard Attenborough) had said, "An eye for an eye will only make the world go blind". There cannot be peace without trust and there cannot be trust without good intentions and humanity. The warmongers may romanticize war but it is always to be remembered that war involves the spilling of the blood of thousands of innocents and it has no end.

However, given the distrust between nations, it only can be said that the greatest challenge and the most difficult fight is the battle against the evils of war. War would always try to rear its ugly head. As nations continue to disbelieve other nations, the only way to prevent a war is to prepare for it. As the former missile scientist and former president of India, Late A.P.J Abdul Kalam had said, "Only strength respects strength". Only strength will deter an adversary.

Sadly, this is the truth. Nations with even defensive intentions and pacifist principles will need to prepare for war more thoroughly to keep the upper hand and to maintain peace. To prevent war we need to prepare for war itself. This along with noble intentions can be the only guarantor for peace.

Conclusion

"War will never cease until babies begin to come into the world with larger cerebrums and smaller adrenal glands."

– **H.L. Mencken**

"War may sometimes be a necessary evil. But no matter how necessary, it is always an evil, never a good. We will not learn how to live together in peace by killing each other's children."

– Jimmy Carter

World War II was not the last war fought. However, since then, the world has been less violent because of the increased sophistication of weapons and the notion of mutually assured destruction. When compared to other periods of human history since the 1950s the costs of war and the damage it can do to the economy have increased greatly and this has kept large scale wars limited.

Having said that, however, it is not that wars have been eliminated or that war will not happen. It is just that it will take a different shape in the form of asymmetric warfare and irregular warfare, where deniability and ability to limit culpability would embolden enemy states to spar at each other without direct confrontation. More than a blitzkrieg we are going to see the slow bleeding and erosion of authority being employed by enemy states to counter stronger opponents. This is the new reality we have begun to recognize. Since 9/11, the battlefield has shifted to new fronts, fought by enemies who shoot and scoot with no real global address or single commanding authority. Every time a central enemy figure has been eliminated a new one has emerged. Challenges and the landscape have been constantly changing. To uproot completely and decisively this hydra headed monster, the world would require the coordinated effort of a number of like-minded nations.

Chapter 19

MARRIAGE AND CHILDREN

How the concepts of marriage and
children could be turned on its head

"By all means, marry. If you get a good wife, you'll become happy; if you get a bad one, you'll become a philosopher."

– *Socrates*

"Marriage is not a noun; it's a verb. It isn't something you get. It's something you do. It's the way you love your partner every day."

– *Barbara De Angelis*

The very definition of marriage has been evolving over the last 50 years. There has never been a more fundamental change in the way we define "marriage", as happening right now. The framework of marriage encapsulated in a legal, emotional and sexual relationship is on the cusp of something new. The contours of marriage are being redrawn as we speak. The first challenge to the time tested mores and customs came in the 1960's and 1970's revolution in the US, where there was a cultural strike at the very foundations of marriage bringing in a revolution that has its resonance even today.

Many people still view marriage through the lens of the 1950s, where the husband was seen as the breadwinner and the wife as a homemaker. This has however undergone a sea change since then. Even many decades later, the widespread cultural changes through the 1960s and 1970s in America, has had wide resonance world-wide.

Most of these changes have their moorings in the economic liberation of women as they began to join the workforce. This has led to greater economic empowerment of women and their ability to exercise their own choices that go beyond the traditional. Nowadays, women are less likely than before to view marriage as crucial to their financial security.

Marriages of the past were **defined by 7 pillars**

- Being married to one person
- Marriage is a heterosexual relationship between a Man and a Woman
- Marriage is meant to be a lifelong bond
- Marriage is a legal contract that is also morally binding
- Marriage meant sexual fidelity to that one partner forever
- Children were born only of wedlock
- Parents provided for the children until they could fend for themselves

Throughout the history of recorded time, marriage was meant to be a kind of economic and social arrangement. It was a kind of a family affair and a contract between families with property, wealth and other inheritances that needed to be taken care of. These were the realities of marriage which went beyond religion or love. The couple to be married had typically little to say in the arrangement. This was mostly true in marriages amongst the economically privileged in society.

Economically challenged communities tended to look at marriage from the practical aspects of daily life rather than personal desires. As civilization grew with its roots in agriculture, families needed more hands on the land to control and exploit it to the fullest extent. These considerations often drove the calculations of marriage.

In modern times with greater economic emancipation, marriage was being looked at from the prism of love and lifelong commitment. However, for women, in particular, who are gaining economic independence, marriage as an institution began to be viewed differently. Non-traditional lifestyles have taken hold and education, career and job have begun to take priority over marriage. Nowadays, many couples have chosen to remain single or have begun cohabitating because of financial instability, debt, and unstable jobs. Americans and Europeans, in particular, view marriage as an out-dated concept and as a piece of paper with no relevance to modern contemporary lifestyle and culture. Marriage in America in the first decade of the century has become the preserve of the educated and middle/rich classes with the other segments unable to gain the financial stability to make it work.

With the fall in wages of men and increasing empowerment of women, women are exercising non-traditional choices. This was compounded by the factor that women are also less willing to marry someone who makes less than them. Sometimes, the marriage involves the reversal of traditional roles. Change is even reflected in Hollywood movies which now portray working women in powerful positions while men stay at home with children. It also depicts couples in successful same-sex unions.

Although marriage is as old as human civilization, it has been in a state of constant flux in recent times. In contemporary times all the 7 pillars of marriage are facing a myriad number of challenges. Concepts such as 'Polyamory', meaning love (amorous) for more than one person (poly), have begun to take hold. 'Polyamory' is a lifestyle where the couple can choose to be in a love with more than one person, sometimes living under one roof.

Marriage is no longer an exclusive preserve of heterosexual people. Marriage has undergone a tectonic shift and is now defined by love and not by gender anymore. Same couple relationships are legal and many a time officially recognized in many countries. The official sanctity gives these marriages legal sanctity, including rights to adoption.

It has become unusual nowadays to think that a couple getting married do it for life. Divorce and separation have become so common that it no more raises any eyebrows and is no more seen as taboo. Many relationships are nothing but 'serial monogamy' where the person marries another and then another after divorcing the previous partner. Adult life is composed of serial co-habitations and some resultant off-springs. The country recording the highest divorce rates is the island nation of Maldives with the highest per capita divorce rates, followed closely by Belarus and Belgium recording to over 65–70% divorce rates. Belgium as a nation has better social security for singles and there is no social stigma towards divorce. While each society has evolved irreligious or religious views on marriage, Belgium women take less than traditional view of married life and are highly educated. Mutually agreed divorces are typically highest in Belgium.

The divorce rates for married couples in the United States stands 5th highest in the world at 50%. Interestingly though in the US, the divorce rate has traditionally been lower for couples during recession, when financial need and investment in equity in the home kept the couple together. Divorces often surge back when the economy recovers and couples then sell their homes and put up divorce papers. New concepts such a 'divorce mediation', 'collaborative divorce', 'divorce counselling' and 'conscious uncoupling' have become more prevalent to cushion the legal battle that ensues divorce proceedings. These concepts help to make divorce a less painful and expensive process and help couples to reconcile to the new reality without a bitter separation battle.

Also, marriage no longer guarantees sexual fidelity. Many marriages that appear monogamous on the surface may have many secret affairs. Marriage partners even negotiate a more fluid type of relationship with outside partners and sexual engagements that do not threaten the emotional moorings of their marriage. Here emotional commitment is the anchoring factor and not sexual exclusivity.

Major cultural shifts have altered perspectives on marriages with many European countries such as Austria, France and Germany, legalizing same-sex unions. In the United States, Australia, Canada and some European countries too, there are altered opinions and acceptance of sex outside of marriage, single parenthood, inter-racial couples, live-in relationships or even individuals who choose not to marry.

It is possible that in the future marriage may be defined by shorter more renewable contracts with a kind of lease period. These may be revisited and renewed depending on how the expectations are met on mutually acceptable terms. This would not just be for heterosexual marriages but also for same-sex marriages. While Polyamory might be commonly or even socially acceptable, even polygamy may be revisited in the legal system.

We may even see more monogamy agreements that are revisited. People may have more open honest conversations about their desires and fantasies.

There may be new visions for the relationship which may be renewed on a regular basis. Sex will be seen as a necessity rather than a sacrosanct factor and will be seen in the light of the health and well-being of the individual rather than through the prism of marriage.

A majority of people from countries of high divorce rates think that marriage is obsolete and as a result, more people are single or divorced and it even suits them to be that way. While marriage may give legal sanction for children to inherit property, the lack of these motivations, may result in the decline of marriage in itself. Couples nowadays commonly have children out of wedlock. Now, there are other mechanisms to transfer property to one's children. Partners may move from marriage to having a 'primary partner', who would be the person one has a deep emotional and spiritual connection with.

With advances in medical science, people are also likely to live longer and this would mean people have time for more than one serious relationship lasting over a 'lifespan'. Hence, the concept of one soulmate would change as humans move along in their social evolutionary process.

The Future of 'Children'

"There are only two lasting bequests we can hope to give our children. One of these is roots, the other, wings."

– Johann Wolfgang von Goethe

"There is stardust in your veins. We are literally, ultimately children of the stars."

– Jocelyn Bell Burnell

Partners nowadays are waiting longer to get married. Some women are freezing their eggs and putting off childbirth. We are going to see more egg donor-ship and surrogacy in the future.

In 1978 the first test-tube baby Louise Brown was born. It was accomplished by in-vitro fertilization or IVF as it is popularly called. Those were the years where religious, ethical, and societal questions dogged these developments and raised controversial questions. Nevertheless, IVF today is commonplace and a natural choice for couples who are having trouble conceiving.

In the future, even if we assume that we are not able to prolong the child bearing age of women, we would still be able to preserve their eggs to allow them to have children later in life. This can be achieved by the freezing the eggs for women before they turn 35. The freezing of the ovarian tissue of women at a young age is another option. These can then be thawed and put back into the women years later.

Sperms have been created by scientists from stem cells and it is theorized that the same can be done for the eggs. All these technologies, both current and future would help women exercise free choice regarding when and how she would like to have a baby.

Pre-Implantation Genetic Diagnosis and Designer Babies

In surveys conducted it was found that a large number of respondents are supportive of a technology called *PGD (Pre-implantation Genetic Diagnostics)* as a means of screening embryos for mental retardation. A smaller percentage of respondents are okay with PGD being used to test for athletic ability, improved intelligence and to know about the anticipated height of the child before selecting it. While these findings may seem controversial to some, it does show the general trend towards greater acceptance of these technologies. IVF, when it was first introduced, did face similar resistance but became mainstream when its benefits became apparent in the following decades. It is presumed PGD would go through a similar bout of resistance and then acceptance.

PGD or "embryo screening" as it is sometimes called involves a process of taking a 3-day embryo and extracting one of its six cells to test for genetic markers. The embryos that pass the tests are then implanted in the woman's womb. The future of PGD may involve less invasive techniques to test embryos and given PGD's benefits, it may well be that in the future IVF may become the norm and the way to conceive. People who conceive normally may well begin to be looked at as gun-slingers and radical risk takers!

Gene editing techniques like *CRISPR/Cas* have begun to spark a revolution in the fields of genetics and cell biology. There was however a moratorium imposed on CRISPR to first sort out the ethical and moral dilemmas to conduct further study of this new technology. This moratorium has however been ignored in some quarters.

CRISPR/Cas is a DNA cutting technology first demonstrated in 2012. The technology acts like a pair of DNA scissors that cuts wherever prompted by a special RNA strand. This causes the cell's DNA repair mechanism to take hold which can be hijacked to disable a gene. The high specificity of the technology, ease of navigation and location of a specific DNA sequence all make this technology very potent.

These kinds of tests, may however, lead to the demand for "designer babies" which raises the usual ethical and moral questions of whether we are trying to play "God". However, no one can deny the benefits of being able to eliminate genetic defects and being able to identify embryos that may develop high risk problems later in life. While "designer babies" do present ethical and moral dilemmas the benefits to humanity are compelling and it is hard to argue against having the ability to make an informed choice when it is possible.

The problem some people argue is that we may soon end up ordering designer babies like we order pizza. There may soon be time when we may order specifics by saying "I'd like a girl with minimum attainable height of five feet 10 inches, to be born with green eyes, with genetic pre-disposal to metabolise food and not gain weight, to be physically well endowed, with light or blonde hair and no chance of breast cancer" and the phrase "bun in the oven" will truly have new meaning! This is well, no different from ordering a pizza, with extra cheese, jalapeno and tomato toppings on a thin crust.

It is also rightly or wrongly feared that PGD could lead to less genetic diversity in our population. In some ways in certain circumstances in the future the lack of genetic diversity could threaten the survivability of the human race. Some say that certain gene types may well have certain appearances and endowments while others may have resistance to certain types of diseases. While it may be alright to design babies to avoid genetically caused development problems and growth issues, it may not be wise to rid the planet of the diversity which has always helped survival and evolution of species. Well, the debate is still on.

Artificial Wombs and Petri Dish Sperm

J B S Haldane, a scientist in from the 1920s had famously predicted that by 2074 more than 70% of the humans would be born out of an artificial womb. While that may be many years away, it is now possible for babies to spend an increasing amount of gestation time outside the human body. Babies today survive despite being born at 22–24 weeks. These were unthinkable even 50 years ago. While women nowadays have surrogacy as an option, a 100 years from now we may well be watching babies grow in artificial wombs hooked up to placenta machines with real-time updates on their progress being fed to the parents.

Movie aficionados may recall the movie the *'Matrix'* where humans were nurtured in an artificial medium in huge glass jars hooked up to life support systems. Well, this piece of science fiction may well cross over from celluloid to medical reality. Gestating a foetus outside a womb or *'Ecto-genesis'* as the process is called may sound like science fiction, but is now becoming reality. Researchers at the Children's hospital in Philadelphia kept a premature lamb foetus alive in bio bags filled with fluid for a whole of four weeks. On the other hand, at the University of Cambridge, a team of scientists sustained an embryo in a petri dish for 13 days (one day shy of the legal limit). The point where these two technologies meet is the point where *'Ectogenesis'* would be born in the future.

It is theorized that in the future 'Ectogenesis' would look like a UV protected glass cabinet which is sustained by an artificial lung, circulated by an artificial heart and filtered by an artificial kidney. Welcome to the world of the *'Biobag'*.

In 2009 researchers at Newcastle University in England had announced the successful creation of human sperm from embryonic tissue. Though they were some limitations in the technology it is only a matter of time before such technologies are able to deliver. Scientists have also begun work on producing eggs and even embryos. Interestingly, even gay men in the future may be able to have children of their own as it is possible to make eggs from male cells. Unfortunately, though, such a technology cannot be applied for lesbian couples as a sperm requires a Y chromosome found only in men. Three parent embryos involving a lesbian couple and a man, however, are a possibility.

While science and technology in these areas continue to develop many moral and ethical questions surrounding this progress needs to be addressed before there is any mainstream acceptance. We can be certain though that the future is sure going to look different from what we are used to now and there are going to be continuing changes as we move ahead. It would be an age when the created becomes the creator!

Chapter 20
SPACE AND COLONIZATION
Space the 'Final Frontier'!

"What is it that makes a man willing to sit up on top of an enormous Roman candle, such as a Redstone, Atlas, Titan or Saturn rocket and wait for someone to light the fuse?"

– Tom Wolfe, the Right Stuff

"Across the sea of space, the stars are other suns."

– Carl Sagan, Cosmos

Theoretical physicist and cosmologist Stephen Hawking has argued passionately for space colonization as a way to save humanity. He had theorised that humans could become extinct in the next thousand years unless, humans colonize space. An interesting concept called *'reverse colonization'* of Earth to restore human civilization in the event of a catastrophe has been proposed by physicist Paul Davies. The colonization of space and a colony in space could act as a 'backup' in the event of a catastrophic occurrence on Earth.

Other than just being a backup, the colonization of space has other immense benefits. The resources in space in terms of material and energy are enormous. So much so that according to differing estimates there are enough resources to support thousands to billions times the population on Earth.

While launching materials from the Earth are prohibitively expensive, bulk materials for colonies could come from *Near-Earth Objects (NEO)* such as the moon or other asteroids. The benefits include lower gravitational forces and no atmospheric drag. Many NEOs contain vast quantities of bulk material and metals. Some of these NEOs even have billions of tonnes of water, ice, hydrocarbons and nitrogen compounds under their drier outer crusts. Farther into space, Jupiter's Trojan asteroids are thought to be rich in ice and other volatiles.

An important part of space colonization would involve the mining of asteroids. The Earth doesn't need to be the only source for materials and water. In space, mining and fuel stations on asteroids could facilitate space travel. It is estimated that using propellant from asteroids for the exploration of moon and Mars could save over a $100 billion. All this lies within the realm of current possibilities.

While building space colonies may seem like a daunting task as they present huge technological and economic challenges, steps towards making it a reality are happening as we speak. The requirement to produce all material and living needs of human life in hostile environments such as outer space is a huge challenge. While living on a space station for few days in conditions with zero gravity has its own set of challenges, creating a sustainable colony that would be self-sufficient would be profoundly taxing. Things we take for granted on Earth will need to be looked at and it would also involve technologies to provide controlled ecological support systems that needed to be wholly developed. There are many more unknown issues which need to be tackled.

The Space Race

"I think we are at the dawn of a new era in commercial space exploration."

– Elon Musk

"As long as we are a single-planet species, we are vulnerable to extinction by a planet-wide catastrophe, natural or self-induced. Once we become a multi-planet species, our chances to live long and prosper will take a huge leap skyward."

– David Grinspoon

In recent time there seems to be a renewed urgency to augment current technologies to be able to reach heavier loads into space. With China racing to get its own space station and the US responding by wanting to quickly put a man on the moon again, we have a new space race brewing. The US wants to use the moon as a base for the future colonization of Mars. With Elon Musk laying his urgency bare in the race to Mars, the race is heating up. The dice has been rolled for the human colonization of space. Elon Musk has gone so far to predict that a ticket to space would cost as little as USD 100k to 500k. In his own words, "Very dependent on volume, but I'm confident moving to Mars (return ticket is free) will one day cost less than $500k & maybe even below $100k. Low enough that most people in advanced economies could sell their home on Earth and move to Mars if they want". Such predictions though premature show the intent and ambition of these space enthusiasts.

Would the colonization be done with new technologies or by augmenting existing ones is to be seen. There is the talk of reusable vehicles, the contours for which have been drawn up by *SpaceX* and others.

Space tourism could be the start of things to come. The company *BlueOrigin* of Jeff Bezos is making great strides towards the goal of space tourism. In 2015, its Flagship suborbital vehicle called *New Shepard* reached the *Karman line* of 100 km above the Earth widely regarded as the edge of outer space. A similar feat was attempted by Richard Branson's *Virgin Galactic*. Although it fell well short of the Karman line, it has brought us one step closer in making space tourism a reality. Eventually, though Jeff Bezos's BlueOrigin wants to go well beyond the Earth's orbit and complete a lunar landing before 2023.

Space Elevators

"We've gotta become the Martians. I'm a Martian – I tell you to become Martians. And we've gotta go to Mars and civilize Mars and build a whole civilization on Mars and then move out, 300 years from now into the universe. And when we do that, we have a chance of living forever."

– Ray Bradbury

One of the ideas that Google was backing early on, was the concept of space elevators. The idea that we could build an elevator from Earth into space and transport anything from the Earth straight up and into space is fascinating. The primary problem with this idea is that the elevator would not even be able to hold up the weight of its own columns. Researchers tried to go around this idea by experimenting with extreme lightweight and strong materials. However, when you really think about it, going into space is a long distance up and away, and no matter how light the material, a column of over 250–500 km is just not happening with our current technologies.

Anti-Gravity

"Man must at all costs overcome the Earth's gravity and have, in reserve, the space at least of the Solar System. All kinds of danger wait for him on the Earth… We have said a great deal about the advantages of migration into space, but not all can be said or even imagined."

– Konstantin Tsiolkovsky, The Aims of Astronautics, 1929.

The concept of *'Anti-gravity'* is very interesting and is worth exploring. This could be the real and only way for humans to truly escape the Earth's gravity and set a base for the colonization of our solar system and beyond. The question is could we truly turn off gravity and just float into space. I am no physicist, but knowing human ingenuity, I would reckon this would happen sometime in the future. I really think this is not an implausible idea. I truly believe this is possible. It is just that at this moment we are limited in our knowledge of the evolution of the universe and are still learning. As new dimensions are explored and as new doors open this would become a reality.

In early Indian texts, there is a mention of *Puspak Viman*. While some speculate that it used mercury and sunlight to propel itself, some others say that it used an electromagnetic force to propel it. I would like to speculate, considering its huge size, what if it actually used anti-gravity in some form?

Quantum Mechanics came into existence in the 1920s to explain experimental results which defied classical explanations. When a single photon was given a choice of passing through one or more experimental slits, the single photon somehow managed to pass through both of them. The photon was thus a wave. Experiments that had been done earlier had suggested that it was a particle. The Quantum Mechanics accommodated for the anomaly even though it defied common sense. Thus the photon was both a wave and a particle.

Einstein's theory of special *relativity* and quantum mechanics though not in obvious conflict, do not gel. While the theory of relativity drew a tight connection between energy and mass, quantum mechanics worked hard to dissolve these tight connections, thus breaking the distinction between a particle and a wave. The idea of the quantum field theory is that the universe was made up of fields in which particles appear as focused ripples, knots, some temporary and countable tightening of things. Particles are just bundles of energy and momentum of fields. All particles are but fields and are born out of fields. These particles are nothing but an epiphenomenon arising from fields.

Incidentally, in the year 2012, Scientists at CERN where the Large Hadron Collider (LHC) is located in the Swiss Alps confirmed the detection of the long-sought 'God particle' or *'Higgs Boson'* particle as it is called. Named after a British scientist Peter Higgs and an Indian origin scientist, Satyendra Nath Bose, the subatomic particle called the 'Higgs Boson' particle has great significance. The work of these two scientists changed the way particle physics has been studied ever since. The 'Higgs Boson' particle is the reason why all matter in the universe exists today. The work done by Bose and Albert Einstein which was later added on by Higgs, led to the discovery of the 'Higgs Boson' particle.

The experiments to prove the existence of the 'God Particle' were carried out at CERN which is spread over two countries over the Swiss-Franco border and is 27 Km long and over 70 metres below the ground. The experiment done at CERN seeks to recreate the conditions of the Big Bang and the origins of the universe as of 14 billion years ago.

Taking off from where we left 'Higgs Bosons' could be described as making up an invisible field of energy through which other particles fly and are slowed by it as it imbues them with mass. While we all know that Newton's law of gravitation states that the gravitational attraction between two masses is directly proportional to the product of the masses and inversely proportional to the square of the distance between them. It goes without saying that the weight of an object on Earth is because of Earth's gravity or the force of the Earth's attraction towards it.

Gravity is just a consequence of masses wrapping around the fabrics of space and time. Just like bowling balls would sit on rubber mats and deform it, masses tended to cause space and time to curve and bend. Interestingly since a particle is nothing but a knot in a field of energy, what if there is an invisible energy which can turn off the forces of this attraction and hence mass. Just like a 'God particle' imbues mass could there be another 'energy field' which negates it and if so, can we turn off gravity in a specified object placed in that field. If this is possible then we have the possibility of anti-gravity which when combined with the concept of a space elevator could be the corridor into space which would be nothing but a field of energy that negates mass. In such a scenario, any mass placed into the field could become weightless and be propelled into space. Science-fiction? Maybe, maybe not. If this happens it would give humans a strong basis to escape Earth and begin to colonize space on a mass scale.

What would happen remains to be seen, but one thing is for sure the race to colonize space is going to be 'Out of this world'!

Chapter 21

SOCIETY AND CULTURE

How technology and economics are going to impact society and culture

> *"The decadent international but individualistic capitalism in the hands of which we found ourselves after the war is not a success. It is not intelligent. It is not beautiful. It is not just. It is not virtuous. And it doesn't deliver the goods."*
>
> – *John Maynard Keynes*

> *"'Capitalism' is a dirty word for many intellectuals, but there are a number of studies showing that open economies and free trade are negatively correlated with genocide and war."*
>
> – *Steven Pinker*

Does economics affect the way society and cultures evolve? Does technology subtly affect the cultural moorings of a society? On cursory inspection the impact of economics and its major role in society and culture may not be apparent. Environmental factors such as geography and climate, and factors such as economics and feasibility have dictated everything from food, clothing and housing in every society. This has resulted in the formation of unique cultures around the world. Economics influences the evolution of society, for example agrarian societies are different from coastal societies and their merchant cultures. While commerce and economics have been moulding and shaping civilization, technological change came in as a subtle influence that then brought about paradigm shifts in society and culture. These subtle influences then challenged traditional ways of doing things. The impact of technology is undeniable.

Why Economics?

"Start with the idea that you can't repeal the laws of economics, even if they are inconvenient."

– *Lawrence Summers*

"Technology can create needs even as it addresses them."

– *Andrew Yang*

When *Coca Cola*, the famous soft drinks giant re-entered India after liberalisation of the Indian economy, Coca Cola's executive head was asked in an interview about how he planned on getting ahead of his competitor *Pepsi*. The executive immediately said that his competition was not Pepsi but *nimbu paani* (lemonade) and tender coconut water sold by small vendors. He was clear that his real competition was from the juice and refreshment vendors at every street corner.

Initially, the prices of both Coke and Pepsi were not competitive when compared to the fresh juices sold by street vendors. 250ml Coke sold for more than a serving of tender coconut water or nimbu paani. However, with time, the cost of Coke and Pepsi when compared to traditional drinks dropped, while traditional natural drinks became more expensive. This initially shot up the demand for soft drinks allowing them early traction in the market. Such is the power of economics. However, as the middle class grew with greater purchasing power, they began to prefer healthier options, even at higher prices. This is why the soft drinks market in India will not reach its expected potential because nowadays most people have begun to prefer natural and healthier options.

World over natural juices and cold pressed juices are becoming more and more popular, even if the costs are steep. With consumer awareness, consumer behaviour shows that people are making healthier choices. Thus we are again going to see the re-emergence of the *juicewalas* who sell at every street corner. This shift in choice was only brought on by economics and changing priorities. The short term cost of quenching thirst is weighed against the long-term healthcare costs of unhealthy drinks, and hence more and more people are making informed choices based on holistic thinking.

The Role of Economics

"I looked back on the roaring Twenties - with its jazz, 'Great Gatsby,' and the pre-Code films - as a party I had somehow managed to miss. After World War Two, I expected something similar, a return to the period after the first war, but when the skirt lengths went down instead of up, I knew we were in big trouble."

– Hugh Hefner

"In human life, economics precedes politics or culture."

– Park Geun-hye

As much as technology impacts society and culture, economics is the *'Tipping factor'*. The role of economics on society is not to be dismissed. It is sometimes the sole deciding factor. However, technology has a pivotal role to play in that it makes the economics possible by driving down the price of goods and services. In this way, new technologies get mainstream acceptance, once the trade-offs are just right.

Once new ways of doing things get into the consciousness of the people, society sees change. The new generation are often the early change agents. Whether it is the advent of radio, television, movies, cable television, direct to home or now OTT ('Over The Top' media streaming) services, these all have impacted and changed society and exposed us to varied cultures. Music cassettes, compact discs, *Mp3*, *walkmans*, iPods, mobile phones, *bluetooth* have over the years altered popular culture and imagination.

White goods, such as refrigerators, dish-washers, washing-machines, microwave ovens, cooking ranges etc. have provided convenience and saved time. These have also enabled the liberation of women who were then able to invest time in their careers and take on paid work. This has then led to gender equality and emancipation of women.

The internet and mobile phone services have caused an information explosion, impacting education, cross cultural exchanges, food habits and dressing styles. It has changed and has empowered millions. Social media and access to news has even helped people in repressed countries to protest and demand freedom. Social and environmental movements have gained traction and have impacted the way we think and behave.

One of the most impactful changes has been brought about by cheap data services provided by mobile carriers. It has democratised information and allowed the marginalised to access information and services. It has also helped governments reach out to people in all corners. Hopefully, this would lead to more equitability and fairness, where no one would be left behind.

In the future, with time, we are going to see the costs of education and healthcare drop. Not only is society going to become more aware of the marginalised, it is also going to get to the point where it will be able to guarantee minimum living standards to all people.

As we begin to enter the 'golden age' of human history, information and knowledge are going to be a commodities while innovation and creativity are going to be highly valued and prized. While the marginalised have been caught in their struggle to fulfil day to day needs, the new age will bring about change and every one would have the opportunity to pursue and realise their full potential. When we enable all, we would be able to tackle all problems facing us and unlock the massive human potential that has been wasted in the past.

While human ingenuity can be trusted to find solutions to our most pressing problems, human nature could prove to be our Achilles heel. However, if guided by wisdom we could prevent our self-destruction and rise above petty differences to achieve prosperity and happiness for all. In this context, wisdom is more important than intelligence.

The Revolutionary Milestones in Societal Evolution

"Human society has dense borders – economic, religious and cultural – inculcated from an early age. We hate change."

– Alejandro Jodorowsky

"We inhabit an obscure planet, in an obscure galaxy, around an obscure sun, but on the other hand, modern human society represents one of the most complex things we know."

– David Christian

Technological progress has changed the economic standards of people around the world for centuries. With better economic prospects and time freed up for activities beyond work, lifestyles have undergone vast changes. Society has been impacted and even cultural aspects have seen adaptations and change.

Agricultural Revolution

When agriculture first took root, civilisation began. Families were nurtured and population began to grow steadily as more tracts of farm land came under the plough. Economic growth corresponded to the increase in population. As larger families worked the land, production increased but this was offset by more mouths to feed. This also meant that per capita growth was stagnant. This revolution was the first agricultural revolution which laid the foundation for civilization.

Industrial Revolution

Once the industrial revolution took hold with mass production, large quantities of goods were produced. As efficiencies increased, goods became affordable. Incremental innovation and newer technologies helped to create big shifts. This contributed greatly in increased production and affordability of the produced goods. In the industrialised and developed countries this helped spawn the middle class, who catalysed this revolution and were at the fore front it. Even still, the vast majority of people worked hard to make ends meet and to put food on the table. The generation of workers during the Industrial revolution toiled long hours to sustain a roof over their head and provide for their families. This generation worked for their needs.

Computer Revolution

However, as technology leapfrogged and with the coming of the electronic and computer age, productivity multiplied and provided workers with much improved wages and quality of life. While electronic and white goods made life easier freeing up time for work and leisure, entire new verticals and industries grew. Computers enhanced productivity and there was a quantum leap in the efficiency of many sectors. Industries were able to manage their production activities, material and human resources in a more efficient and effective manner. Across industries higher wages and more affordable products meant workers were no more working for sustenance but were now working for a standard of living. This was the generation of the IT (Information Technology) revolution.

MIS (Mobile, Internet and Social) Revolution

Gradually, with the internet age and the advent of the smart phone a new era germinated. The great increases in productivity enabled by explosive innovation over decades meant that this generation would come into an era where they would be paid remuneration that would be beyond just sustenance or standard of living. This generation is in the midst of the social revolution where work life balance is central and priorities beyond money matter. This generation has different needs. It is hyper connected and every moment in one's life is shared, celebrated and broadcast, making each person a mini celebrity. This generation is sensitive to the image they showcase and have the luxury of work life balance and great pay. It is a generation that has numerous choices and styles of living they could choose from and afford. This revolution has showered the generation with a surfeit of life choices.

AI Revolution

The next generation though is going to be born in the midst of the AI (Artificial Intelligence) revolution where most mundane work is going to be taken over by machines. Work weeks are going to shrink further and may be as little as 12 hours a week. This generation is going to live with not just a high standard of living but are also going to be able to pursue their passions beyond work. On the bedrock of decades of work done by the previous generations, this generation is going to enjoy the complete fulfilment of their dreams. It is going to be a future where time, money, tools and opportunity are all going to come together opening the door to phenomenal possibilities The generation is going to be able to live life on its terms and achieve great pinnacles. Their limitations would only be personal and they wouldn't need to overcome much of the economic and social barriers that have held back generations of the past. To rephrase the words of Oprah Winfrey it would be an age where truly one's attitude would be the only limitation to one's altitude.

Conclusion

No society or culture around the world has been left untouched by technological change. Within generations, societies adapt to the change. Cultures gradually soften and bend to accommodate the new needs, desires and aspirations of the younger generation. However resistant one may be, change is ultimately inevitable. It is easier for societies to accommodate and move along with the change than to try and repel it.

Often change is for the better, and if not, society will outgrow that change anyway. So there is nothing to doubt or fear. Embrace and ride the waves of change coming your way. With a little flexibility and fluid thinking, one would be able to adapt and reap benefits of these waves. So raise your sails and enjoy the ride!

Chapter 22
BALANCE OF WORLD POWERS

How there are going to be dramatic shifts in the balance of world power this century

"He who has great power should use it lightly."

– Lucius Annaeus Seneca

"Our great power does not mean we can do whatever we want whenever we want, nor should we assume we have all the wisdom and knowledge necessary to succeed."

– John McCain

Before we try to understand the shifts in world power that this century would bring, we need to understand what we mean by world power. World power has three major dimensions. These are 'economic power', 'military power' and then there is 'soft power'. It is expected that in the current century, the world powers would rebalance themselves and there would be a shift in the center of power with the rise of new players.

The Asian Century

"Before we acquire great power we must acquire wisdom to use it well."

– Ralph Waldo Emerson

"Leadership is about vision and responsibility, not power."

– Seth Berkley

This century would see a reset at the table where world powers meet and discuss. It may not be known to many people that for thousands of years until late 1700s, the centre of world power, its *GDP* and soft power lay in Asia, with the two Asian giants, namely China and India. Together, they constituted more than 60% of the world's GDP. They were also the power centres of education, culture and the arts. From the late 1700s however, over the next 250 years, there was a decline in their shares of GDP and world power, to the point that they became marginal players. Europe and then America rose through rapid industrialisation, while Asia stood exploited for resources that rapidly dwindled as the Europeans took over the world.

However this century, China and India have begun to play catch up in turbo mode. China with a head start on reforms has shed its communist leanings to embrace capitalism and has risen phenomenally over the last 4 decades. India is 15 years behind China and has begun to pull up its socks. It has transformed its mixed economy with economic liberalization and market reform. It is expected that by mid-century China, India and United States would be the three largest economies in the world. The shift of the economic power centre to Asia would be complete.

Beyond China and India there are more, well, surprises. Indonesia could become the fourth largest economy in the world while Mexico and Turkey could grow bigger than Germany and France. Other rising powers include Saudi Arabia, Nigeria, Egypt, Pakistan, Iran, Philippines and Vietnam. All of these nations would rank higher and would be larger economies than Italy and other smaller European nations.

Since the end of World War II, the US has been a dominant player in the international system and was central in creating new international organisations such as the United Nations, NATO, IMF and the World Bank. American diplomacy spearheaded agreements on trade, climate change, arms control and it also assured regional and world security. Americans were always at the forefront formulating and directing the "rule based international order" and they stood by guaranteeing the same. Moreover, the United States of America took it upon itself to maintain a level of sanctity with respect to international agreements and order.

Since 2016, the US has upheld a more introspective policy under its new President and has also slowly begun to withdraw from its international engagements. This has left a vacuum. Russia and more so China, have begun to fill this vacuum and these nations have become more assertive and visible. It is not just these two major powers but as America's global role has been diminishing, a growing number of countries have begun asserting their independent views and these countries are taking up influential roles in the regional and economic playbooks of their neighbourhood.

However, having made these observations, it is not that Asia's rise is inevitable. If the drivers that are fuelling their rise could falter, their rise in power and domination could also deflate. China could for instance, go into a political meltdown, similar to its last great empire. The Indian economy may face problems from the naxal unrest, communal problems peppered with caste based friction and uprising. Japan may continue to shrink with its ageing population. However, barring any such setback, Asia will see an inevitable rise and the shift of power from west to east is bound to happen.

Economic Power

"Power is like being a lady… if you have to tell people you are, you aren't."

– Margaret Thatcher

"Power is the great aphrodisiac."

– Henry Kissinger

By 2050, the three largest economies in the world would be China, India and United States. China as of 2019 had already overtaken the United States with a higher *GDP (Gross Domestic Product)* in terms of *PPP (Purchasing Power Parity)*. It is estimated that by the year 2030 India would overtake the United States and become the second largest economy in PPP terms.

These projections may surprise many who have been blind-sided by the rapid development of China and by those who have viewed India as a backwater in the past. However, this would not come as a surprise to those who have been following these two nations closely. It is only to be expected that with their large populations these nations would achieve higher GDPs.

GDP is a product of Labour, Capital and *'Total factor productivity'* or *TFP*. Given that these nations are endowed with large number of human resources, even with a small injection of capital, local or foreign, these nations could realise their potential in a short period of time. The TFP which represents the contribution of innovation and technology to the GDP would also accelerate the rise of China and India. With respect to the growth of GDP, other countries like Indonesia, Brazil and Turkey are also likely to figure in the Top 10 nations by the year 2050.

The rise of these nations would cause huge shifts in world trade and lead to the formation of new trading partnerships and blocs. This would lead to a new world order as economic power would facilitate military power, influence and reach. A multi polar world with new power centres would be created. It needs to be seen if this change and replacement of the old world order with a new one, leads to more prosperity or conflict in the future. The world would hope for the peaceful rise of nations, especially China. As Seneca had wisely said, *"He who has great power should use it lightly."*

Military Power

"I hope our wisdom will grow with our power, and teach us, that the less we use our power the greater it will be."

– Thomas Jefferson

"A great power has to have the discipline not only to go when necessary but to know when not to go. Getting involved in ethnic, religious civil wars is a recipe for disaster."

– John Kasich

While the economic rise of China and India is inevitable, the United States, in the early part of this century, would still dominate the world in terms of its military power. However, as China is focussed on becoming a world power, it would provide a stiff challenge to the US and assert its position wherever and whenever possible.

It is also to be understood that the huge influence of the US military-industrial complex creates undue influence on US foreign policy and distorts it. It sometimes forces the United States to act in ways which are detrimental to it in the long term, which detracts from its moral authority. An example of this is the war to oust Saddam Hussein based on questionable evidence of *WMDs (Weapons of Mass Destruction)* which were actually never found. Military force has often been the most preferred choice and has been higher up in the deck of cards that the US wishes to play. It has mostly marginalised diplomacy and other instruments that would have won it more legitimacy and world support. The Trump Presidency has been inward looking, the US military has grown while its diplomatic force has shrunk further, thus fuelling this failing approach. The old adage that "if the only tool you have is a hammer, then every problem becomes a nail" rings true in this context.

China has made its entrance as a formidable power. This along with its new assertiveness and the fading legitimacy of the United States has led nations under the security guarantee of the United States to ponder. It is expected that these nations that are dependent on the United States for their security would need to strengthen their own military and take matters into their own hands. This would create a multipolar world with a number of power centres trying to jostle and outdo one and other. In other words, we are entering an uncertain era where new powers and new power alignments would begin to act independently, which may add to the disorderliness of the world affairs. The 'power vacuums' left by the US are likely to be filled by China and others.

With its economic clout, China is likely to snare a number of smaller countries into its net with projects such as the now infamous 'lend-indebt-snare' strategy. This helps the 'Old Imperial Nation' to buy the ability to form military bases through influence strategies and also establish points of influence to project power around the world. It is also possible that if China does not rise peacefully, a war with one or the other major military power would be inevitable.

While in the new future, China may not be able to match the United States militarily, no one should underestimate the resolve of the Chinese leadership that seeks to get China back to its glorious past. China blinded by past humiliations and with the pride of its magnificent past could be a military threat to a number of other nations. In this century, the United States and China are likely to jostle, push and shove as the United States begins to accommodate the rise of China. Not for nothing does China call itself the

'Middle kingdom'. It truly considers itself the 'Center of the world' and an indispensible world power.

Soft Power

"Power is of two kinds. One is obtained by the fear of punishment and the other by acts of love. Power based on love is a thousand times more effective and permanent then the one derived from fear of punishment."

– Mahatma Gandhi

"We thought, because we had power, we had wisdom."

– Stephen Vincent Benet

Hu Shih, the Chinese philosopher and diplomat once said "India conquered and dominated China culturally for 20 centuries without ever having to send a single soldier across her border". This just highlights that no country really needs to dominate another militarily or economically. It is possible to be an influential soft power without the need to coerce or bribe.

'Power' is the ability to obtain outcomes one would desire. This can be accomplished by 'carrots' using one's economic power, through 'sticks' using one's military power or by using the powers of persuasion and one's attraction. While the former two are components of hard power the last is what soft power is all about. It is the ability to obtain favourable outcomes without payment or use of force.

Rather than thinking of soft power as power over others, one should look at it as power to accomplish anything by taking others along with you. The power of persuasion that convinces and influences many to move along with you is a vital asset and this is what soft power can accomplish. In the last century, Great Britain and the United States wielded their soft power to great effect by being the centre of the global movie and music industries. Along with the advantage of the English language, power over the media the cultural appeal of liberty and their philosophies about freedom, they have been able to fuel dreams which have consequent influence and charm. The early success of capitalism along with the promise of the life of freedom and liberty had its magical effects which even led to the demise of the once powerful U.S.S.R.

The United States overcame the U.S.S.R through the appeal and the benefits of capitalism, with its subtle projection of the good life, promise of liberty and life with freedom.

Such is the power of 'Soft-power'. In the age of media and the power of its other forms such as social media, nations can be appealing because of their soft power and thus they need not necessarily wield the stick or carrot to have their way.

Conclusion

"The attempt to combine wisdom and power has only rarely been successful and then only for a short while."

– Albert Einstein

"What it lies in our power to do, it lies in our power not to do."

– Aristotle

"It is certain, in any case, that ignorance, allied with power, is the most ferocious enemy justice can have."

– James Baldwin

A nation's ability to combine the three powers described namely, economic, military and soft powers determines its true power. Sometimes empowering others helps us to attain common goals. The efficacy of the nation to achieve a successful strategy hinges on its ability to combine these three dimensions of power seamlessly.

At the very foundation of soft power is the subtle appeal of a nation's popular culture and its ability to connect with nations and their people across the world. While 'Soft power' alone may not be able to solve many problems, its influence should not be underestimated. Broad acceptance and ability to forge common ground would be needed among nations to accomplish many of the goals to make this world a better place for everyone. Let us hope that common sense would prevail and more world powers use persuasion and soft power rather than forcing their way with their economic and military might.

Chapter 23
'CREATIVE DESTRUCTION' AT ITS BEST

Creative destruction is the upheaval that technology would un-lease in the job market 'destroying' and 'creating' a new wave of jobs and new verticals

"The best way to drive your future is to be in the driver's seat and create it."

– Author

"Change is the only constant, it comes in waves like the waves in an ocean, the best way to deal with it is to surf it, not fight it."

– Author

In the year 1977, the Tatas of India proposed to the government the complete computerisation of the tax administration system. This was rejected by the then finance minister of India, Charan Singh, and India missed a golden opportunity to implement a fully computerised tax administration system way back in the late '70s.

This is chronicled in a book by management strategist Shashank Shah in his book titled 'The Tata Group: From Torchbearers to Trailblazers'. The book was released on November 29, 2018, to coincide with the completion of 150 years of The Tata Group and the death anniversary of JRD Tata.

In 1977, the then Prime Minister of India, Ms. Indira Gandhi, and her government had rejected computerisation and reasoned that introduction of computers would lead to the loss of jobs and cause mass unemployment. It is ironic that years later, the introduction of computers and creation of the software industry created more high paying jobs and swelled the middle class of this vast country than any other sector. This just goes to show how simplistic straight line thinking cannot visualise the future.

While advancement in technology over the last century has caused a quantum jump in productivity, it has caused the creative destruction of the job market. This in turn left many of those people who did not upgrade their skill, stranded in the job market. Yet it did provide great opportunity to people who adapted to the change and learnt new skills. These people thrived with the change.

These changes have caused groups of people to be left behind, but it did create a path to riches for those others who were willing to adapt. These vagaries have sometimes been responsible for transformation of society. They have also caused political upheaval, as witnessed by the unexpected election of Mr. Donald Trump as the 45th President of the United States. Trump was powered by disaffected voters who were unhappy with the job losses and lack of opportunity in what is called the "rust belt" of the United States of America. This is where the manufacturing jobs that once held roost disappeared.

These narrations go to show the complex dynamics of how technology impacts society and how changes cannot be easily foretold. The future could unravel many dimensions. Often, the unravelling may not be in the manner that people may have expected it to be.

Job Losses

Society has a love-hate relationship with technology. People usually do not embrace or adapt to changes in technology, as any change causes a shift in the current equilibrium, which requires them to get out of their comfort zone. Any shift that may affect the stability of their lives and society in general is not welcome, as all are happy in their state of inertia. If technology needs to be adopted it needs to be compelling. There are always places in the world where fear of job losses and possibility of unintended consequences can cause governments to slow or forestall the adoption of new technology.

Examples of these abound. The greatest challenge faced by companies that create new technologies is in bringing about a behavioural change and roping in early adopters who are willing to adapt and try these innovations. Some companies and their products/services are so compelling that they achieve this more easily than the others. One of the most successful start-ups of our times, Uber, is one such example. It was ultimately a game changer. The business model and mobile app were widely well received. The market then was made up of taxi operators who owned a fleet of taxis or many individual taxi drivers operating from a variety of stands. Uber thrived as the taxi drivers did not have to scout to be hired and the consumers could hail them on their phone. The local governments then intervened and brought in regulations and protectionist policies. These spokes that were brought in attempted to hinder the smooth functioning of the new entity, and to protect and maintain the old guard. Slowly however, those who were capable of change embraced it, while the others retired or changed their working style and ways of doing business. These are examples of how friction is caused by new entrants and how stakeholders in the mix adapt or stall. While some gain advantage and some are disadvantaged. Change remains the only constant as creative destruction unfolds.

The most feared change that comes with new technology and with creative destruction is job losses. There was a time before computers, in the '50s, '60s and '70s of the last century, when the knowledge of 'short hand' (an abbreviated symbolic writing method used by secretaries to jot notes and write down minutes of meetings) and 'typing' were highly valued skills. Those skilled in them fetched high paying jobs. There was a time when a large number of people with typing and short-hand skills headed to Mumbai via trains from many other parts of India. The trains were popular and were nicknamed 'Typist Gadis' or carriages that carried typists. Being a stenographer was a sought after skill and companies headquartered in Mumbai offered great opportunities for people with such skill sets.

Nowadays, with the winds change, short-hand and stenography are dying skills that are both unnecessary and obsolete. Stenographers are a dying breed. Technology has progressed to such an extent that even 'personal secretaries' who were themselves a dwindling breed, would soon be replaced by 'personal assistants' programmes. These would be programmes that run on AI (Artificial intelligence) and do everything from scheduling and booking itinerary to finding one a cab. While these changes have improved productivity they have eliminated and will eliminate low level jobs in the future. These will cause disruptions that will impact society.

India in the first two decades of this century, rode the wave of outsourced software services, *BPOs (Business Process Outsourcing)*, *KPOs (Knowledge Process Outsourcing)* and the low skilled call center services. These created millions of jobs for the masses in India but also as technology progresses in leaps and bounds, these very jobs could see a decline in the future. The low level functions in these jobs could soon be automated and replaced.

As India has been moving up the value chain, we have seen the migration of call center jobs to countries such as Phillippines. New technologies in the horizon would soon cause new kinds of disruptions in the job market for call center employees. 'Chat bots' and programs that mimic call center executives driven by AI (Artificial Intelligence) would soon replace majority of the jobs in the call centre industry. These would not only cut costs for companies in their customer service operations, but result in job losses starting at the lowest levels of call center operations.

It is not just call center employees who would be affected, but software professionals whose jobs could be 'Automated' would also find themselves facing job losses as companies automate mid-managerial level jobs to bring in cost efficiencies and shrink payrolls.

While software outsourcing companies are set to see Increase in profitability as they eliminate payroll, many jobs in this industry would become redundant. New skills and fresh hands would replace those with old skills and jaded experience. These job losses are not just confined to developing countries but developed countries will also face their own set of challenges.

Changes in the Developed World

While one may be lulled into thinking that job losses would only occur in developing countries, this thought is far from the truth. In the developed world, while blue collar jobs were the early casualties of outsourced manufacturing which caused extreme job losses, people often believed that jobs such as truck driving could not be outsourced and were secure. There are an estimated 3.5 million professional truck drivers in the United States of America. Annually, all these drivers combined drive about 400 billion miles and haul more than 10 billion tons of freight. The truth is that while blue collar jobs in manufacturing did get outsourced, they also got automated as machines began to take over the load of manufacturing. During this entire period, people who were truck drivers or cab drivers (though they performed by the lowest rugs in industrialised societies) had secure jobs.

Now, it is believed that by the year 2025 we are going to see a revolution with driverless cars/trucks and drone delivery mechanisms which are going to begin to eliminate jobs for people who are truck drivers, cab drivers and delivery boys. Jobs that were once thought to be safe in the developed world are going to be made redundant by technology and would be under threat. Low level jobs including those created by companies like Uber would also come under threat. It is estimated that there are over a quarter million taxi drivers in the US and at any point of time, there are more than three quarter million drivers that are driving for Uber alone. In the future, the number of jobs in the trucking and travel industry will shrink or be eliminated.

Even e-commerce and food delivery businesses would be impacted if drone deliveries come into vogue. The only thing that stops drone deliveries from being main stream is really a regulatory nod. Once the skies are opened for drone deliveries and proper collision prevention and sky regulation mechanisms are evolved, it would cease to be some science fiction and turn into reality. It would be happening in the skies above us as we go about our day to day work.

Change on an Industrial Scale

With solar power, electric vehicles, autonomous vehicles, and transport for hire coming of age, we are going to see tectonic shifts in the industry. This is going to create and destroy jobs on an unprecedented scale. Large industries as we know it are going to disappear and new larger ones are going to appear. New skill sets are going to be required, while older skill sets would be redundant and useless.

Insurance Industry

The Insurance industry is going to see the disappearance of entire verticals. Car insurance is going to be the first casualty. With AVs 'Autonomous Vehicles' becoming the norm we are going to see a drastic fall in accident and accident claims, making the car insurance industry shrink and potentially even disappear.

Oil and Coal Industry

With the coming of EVs (Electric Vehicles) the entire oil industrial behemoth is going to totter towards extinction. The growth of alternative energy sources such as solar power is going to accelerate this demise. These technologies are becoming economically viable and within a couple of decades, (to be charitable with time), we are going to see the death of the oil and coal industries. The epitaph of the coal industry is going to be written as the solar power technologies begin to eclipse the viability of coal as a power source. The oil industry is going to crash and burn with the EVs shooting down this high flying industry that has powered the modern era. We will soon witness the death of the 'carbon economy' as we know it.

Car Industry

We may see new players emerging in the automotive industry with EVs and automated vehicles taking over. Unless, the entrenched players make a change they may become extinct. Further, we may notice that fewer cars would be needed as the utilization factor of cars goes up with shared services such as Uber. Also, with the coming of EVs (Electric Vehicles) which undergo little wear and tear, people would not be looking to replace their cars ever so often.

Further Depths to Which Technology Can Create Change

For long it was also believed that only routine jobs like factory jobs which were based on a set of rules or tasks could be automated or computerized. Now, with advancement in technologies it has dawned on the public that even jobs that required personnel in a less predictable environment such as driving cars could face the heat of technological change.

However, it was always assumed that professions that were wrapped in semantics and language could never be computerized. Well, that is also about to change with advances in artificial intelligence and technology. The routine task of sifting through documents and looking for specific information or relevant paragraphs has become redundant. This has begun to make an impact on a number of professions that were considered safe from both outsourcing and automation. Though these changes would begin to change the way things are done over next few years, the lull before the storm is being endured with bated breathe.

Impact on the Legal Profession

Many young Americans have nurtured their dreams of becoming lawyers and ultimately rising to the top as partner in a law firm. Fed on a steady diet of onscreen action and drama of legal professionals, right from *'L.A. Law', 'Law & Order', 'The Practice', 'Boston Legal'* to later shows such as *'The Good Wife'* and *'Suits'*, the glamour of life in the legal profession has drawn many youngsters to pick up law as their profession of choice. Many flock to law school taking hundreds of thousands of dollars in educational loans with the hope of a great life. The dream seemed all set, until…

Enter Watson, no not the one in *'Sherlock Holmes'*, this one is by IBM. Named after the first CEO of IBM, Thomas J. Watson, Watson is a super computer that processes at the phenomenal rate of 80 teraflops per second that is 80 trillion floating points per second. It can process 500 Gigabytes or an equivalent of a million books per second. It can replicate the ability of a high level human brain to answer questions. It has access to servers that have a combined storage of hundreds of millions of pages and processes information taking into consideration over six million logical rules. Well, now you may think this is a big machine that is bigger than a Space Shuttle. Sorry to blow your bubble, all it needs is a space as small as a room that can accommodate a dozen refrigerators! In recent times the space required to house IBM Watson as a cloud delivered, enterprise ready solution has shrunk further and it can be accommodated in a space as small as three stacked pizza boxes. Moreover, its performance is said to have improved 23 times over.

When *'Deep Blue'*, the AI machine from IBM defeated Garry Kasparov in 1997, the true power of harnessing 'Artificial intelligence' had begun to catch the attention and imagination of the public. This got the IBM researchers thinking and led to the creation of *IBM's 'Watson'*. The computer system was initially developed for the quiz show *Jeopardy*. It went on to compete and in 2011, it won against legendary champions of the show to win the first place and $1 million cash prize.

Application of Watson's cognitive computing ability has vast applications. Right from law to medicine, the field is entirely open. Operations on huge volumes of unstructured data are possible and are backed by complex analytics and deep mining abilities. In the month of May of the year 2016, an Ohio based firm called *BakerHosteler* signed up for a legal 'expert system' based on the abilities of this super computer. That system had the ability to mine data from about a billion text documents, analyse the contents and provide pin point answers to complex questions in a jiffy. Combined with Natural Language Processing it could respond to questions related to law by translating any legal documents in normal spoken or written English language.

With IBM Watson gradually taking over the lawyer's profession and with its ability to deliver results with over 90% accuracy compared to just 70% for humans, it may not be long before the payrolls at law firms shrink. Freshly minted young American graduates in law are not getting jobs because with IBM Watson one could get legal advice within seconds from anywhere, at any place and at any time. With these possibilities, in the future, entry level legal advice would be available for cheap and would be available on call. There would be less need for entry level lawyers in law firms. It is estimated that any research or due diligence can be entirely done by such a system and a study report suggests that this would reduce the need of entry level lawyers by 24%.

Impact on the Field of Medicine

IBM Watson is also set to revolutionise medicine. It is being used in the field of Oncology to detect Cancer and in many instances they have found it to be more accurate that doctors and nurses. In the year 2016, *Manipal Hospitals* of India, launched IBM Watson for cancer patients to be guided to pick cancer care options. In fact, Manipal Hospitals offers this technology to patients online through their website and is the first in the world to do so. All these changes herald a new beginning in the field of medicine. Though there are numerous challenges in the ultimate delivery of results, this could completely overhaul the system. The die for change has been cast and it is only a matter of time before technologies would start providing answers.

IBM Watson has wide applications right from financial service industry, education, weather forecasting, to water conservation. The future would be vastly different once IBM Watson and its successors/competitors come through with their complete potential suite of applications. Well, is this just the lull before the storm?

Creative Destruction

"Remember today is the tomorrow you worried about yesterday."

– Dale Carnegie

"But there is always creative destruction in markets: There are always new winners taking the place of those that are. So if you look at the market's surface it may appear flat, but there's always huge turbulence taking place within."

– Kerr Neilson

Technology, the economy and the job market will always be in a state of flux. 'Creative destruction' is a given and is essential for the progress of mankind. The only questions that remains are "Are you aware of it?" and "Are you prepared for it?" There is nothing that is going to stop the creation and destruction of industries and maybe entire verticals. 'Fluid thinking' and 'agile adaptations' are the essential attributes for survival and success of individuals in this new future. Those who are cushy in their comfort zones and those who cling on to the past and have fixed ways of doing things are going to find themselves left behind as sweeping changes bring about the need to adapt to the changing shape of the future. Those who do adapt will thrive, while the rest will be left with nothing but nostalgia!

Chapter 24
THE FUTURE CZARS OF TECHNOLOGY

Who are the ones who are going to dominate
the technological world of the future?

"The best way to predict the future is to invent it."

– Anonymous

"Here's to the crazy ones. The misfits. The rebels. The troublemakers. The round pegs in the square holes. The ones who see things differently. They're not fond of rules. And they have no respect for the status quo. You can quote them, disagree with them, glorify or vilify them. About the only thing you can't do is ignore them. Because they change things. They push the human race forward. And while some may see them as the crazy ones, we see genius. Because the people who are crazy enough to think they can change the world, are the ones who do."

– Steve Jobs

Turning the Titanic on a Dime

In the late 1990s Bill Gates the co-founder and then CEO of Microsoft took a decision that was meant to change the direction and the very assumption on which Microsoft was founded. Microsoft believed in and only built personal computing software with a standalone operating system on standalone devices. With the coming of the internet Microsoft faced new and complicated challenges and had to make massive adjustments to its assumptions and do so quickly.

Even though, Bill Gates was slow to recognise the potential of the Internet and the power of the web, he soon caught the drift and his decision and the resultant change in direction was swift to say the least. Initially, Bill Gates thought the internet was just a precursor to some sort of elaborate, multidimensional information superhighway. Microsoft's change in direction was akin to "Turning around the Titanic on a dime". Bill Gates not only woke up to the potential of the Internet, he embraced *Java* which was the cross platform language that Microsoft was averse to until then. Gates adopted a strategy of "Embrace and Extend" as he entered into a licensing arrangement with *Sun Microsystems* (The company that originally created Java). The letter of intent to license Java to Microsoft was signed on December 7th 1995. The letter of intent was in the words of Bill Gates "a kind of famous day". In a keynote he recalled the Japanese had attacked Pearl Harbour on this day. The attack had led to the entry of a reluctant United States into World War II. The Japanese Admiral Yamamoto had then famously observed he feared, "They had awakened a sleeping giant".

Microsoft was indeed a giant and it had awakened. Microsoft also went on to take on *Netscape* in all its fury. Netscape was then the famous browser or window to the internet that had threatened to destroy Microsoft. With the dawn of the internet and the consequent 'paradigm shift' in the market, Netscape had seized the resultant opportunity and had challenged the incumbent. The potential of the browser as a platform for software applications represented an undeniable threat to the dominance of Microsoft Windows Operating System. Marc Andreesen, the founder of Netscape was called the "Next Bill Gates". He demonstrated that the Web was becoming the much talked about and anticipated 'Information Superhighway' and Netscape was the *window* to it and all the riches it could bring. Andreesen boasted that Netscape would reduce Windows to a mundane set of poorly debugged device drivers. In its IPO Netscape was valued more than the defence behemoth *General Dynamics*. It had taken General Dynamics 43 years to be worth about $2.7 Billion then. It took Netscape about a minute to do the same. Netscape seemed to have the future within sight and was set to rise to the top of the pecking order.

Well, not so fast. A determined Bill Gates, suitably awakened, changed the course of his giant company and went after the potential of the internet.

The rest as they say, is history. Netscape slowly bled and was reduced to a pale shadow of its former self. Microsoft bundled the *Microsoft explorer* (its browser) into its operating system to elbow out Netscape. Bill Gates declared Microsoft was "hard-core about the internet". Microsoft Explorer was improved, made faster and offered online browsing for free.

Bill Gates was then dragged into an anti-trust investigation which sought to break the monopoly and positional advantage that Microsoft enjoyed. Bill Gates fought back passionately and side stepped any government mandate to break up Microsoft. Microsoft prevailed in the storm, as the behemoth survived and thrived.

Such is the power of a leader who seizes an opportunity. It also goes to show how taking tough decisions and being decisive makes for a leader who stands out.

The Difference Great Leaders Bring

"At the most basic level, an economy grows... whenever people take resources and rearrange them in a way that makes them more valuable."

– Paul Romer

"As the new endogenous growth theory suggests, TFP growth is closely related to accumulation of the intangible capitals, such as human capital and research and development."

– Toshihiko Fukui

There is a **Classical model** of economics called the *Malthusian model*. **Propounded by Thomas Robert Malthus, the theory states** that with technological progress and land expansion, the production of labour output would increase. However, with higher population growth and an increase in labour workface, the labour input faces diminishing marginal returns. Here the additional output of the workforce declines to zero as it is balanced out by the population increase. This implies that with time we would have a larger population but not a richer population. As its prognosis was to say the least, depressing, this theory was labelled as the 'Dismal Science'.

The *Neo-Classical model* of economics on the other hand tries to find the 'steady state rate of growth' that occurs when the output-to-capital ratio remains constant. Here at the steady state the capital per worker and the output per worker grow at the same rate.

Economist Paul Romer came up with another theory called the *Endogenous growth model* for which he won the Nobel prize in 2018. The Endogenous growth model lays emphasis on knowledge and human capital. The capital spent on R&D and human resources leads to innovation and with such innovation, societies would leap frog ahead and create wealth. In this model because of the R&D and innovations, the diminishing marginal returns to capital do not set in.

It is here that we need to emphasize that a great leader can channel the resources of their company towards innovation, to create unlimited wealth riding on technology. Hence, a great leader is pivotal to creating great wealth and consequently human progress. For this reason it is hoped that going forward we would see many such great visionaries and leaders.

Attributes of a Good Future Leader

"You can't connect the dots looking forward; you can only connect them looking backwards. So you have to trust that the dots will somehow connect in your future. You have to trust in something – your gut, destiny, life, karma, whatever. This approach has never let me down, and it has made all the difference in my life."

– Steve Jobs

"The only place where success comes before work is in the dictionary."

– Vidal Sassoon

Future leaders in technology need to have deep passion and commitment to their jobs. They often see a future, a vision and believe in it through ups and downs. The future 'Czars of technology' would need to be visionaries with unwavering faith in their ideas. While they would get great support from investors looking to join in on the profits, they would need to go through many prolonged difficult periods while establishing their companies. The ability to drive through the numerous obstacles and yet keep faith would be critical. For instance, it took the company Amazon approximately more than 14 years or 58 quarters to be precise, before it could turn a profit. All along, the company CEO and now richest man in the world Jeff Bezos kept faith and kept prodding along.

The founders not only need to keep the faith, the founder would also need to be able to convince their investors and employees to have faith to keep the flock together. These leaders need to be inspirational and magnetic. It was said that Steve Jobs of Apple was able to distort the reality around him when he spoke about his ideas for future products. He not only developed a cult following in the company but also with the customers. This allowed some of his products to fail without failing the company, which then went on to launch blockbuster products.

In more recent times, Elon Musk has caught the imagination of investors who have placed faith in him to transform the auto market with his electric vehicles or EVs. The aura around him has allowed him to stay afloat despite his early struggles and setbacks.

All these instances point to the fact that the future lies in the deep faith that technological leaders have in what they do and their ability to convince stakeholders of their vision. With motivations beyond money, the truly successful visionaries are often highly self-driven and consumed by passion. They yearn to see their creations become a reality and will go through any hardship to make their visions a reality.

We hope that the tribe of such people grow to bring mankind the best of technologies and a bright future.

Chapter 25

FUTURE OF 'WEALTH'

What will 'Future Wealth' look like?

"Empty pockets never held anyone back. Only empty heads and empty hearts can do that."

— Norman Vincent Peale

"Wealth is the ability to fully experience life."

— Henry David Thoreau

Early Wealth

The early wealth of humans who were 'hunters and gatherers' was related to the environment in which they functioned. As they began to rear cattle, horses, hens, goats and sheep, wealth began to be measured by the number of cattle they owned. The early humans were nomadic and wandered from pasture to pasture feeding and taking care of their domesticated stock. Wealth in those times came to be measured by the count of domesticated stock in one's possession. It is pretty interesting to note that even today in many parts of the world cattle wealth is taken as a serious barometer of personal wealth and status. In parts of Africa and India, a person is introduced as an owner of such and such number of domesticated cattle. In fact, in this day and age the practice of gifting a set number of cattle as dowry in marriage still prevails in quite a few places.

The advent of agriculture spurred human settlements and land soon became a resource. Land was fenced and ownership of land and homes became possible and necessary. Wealth began to be measured in terms of acres of land held along with the count of domesticated livestock. Land even today holds great importance and many people including the likes of the current richest man in the world today, Jeff Bezos, all hold large tracts of land. In fact, Bezos is rumoured to have one of the largest land holdings in the US to his name. Real estate in the form of land has and will always be a kind of wealth that humans value because it is a piece of this limited earth. However, its value relative to other forms of wealth has and will continue to undergo sea changes moving it up and often down the pecking order in comparison to newer forms of wealth.

Gold and the Beginning of 'Modern Banking'

Gold has been a wealth barometer for some centuries now and it simultaneously took root as a currency across many different civilizations, as gold began to be used in barter and trade. As a currency it is highly valued for its consistent appearance, aesthetic qualities, rarity, durability and malleability. Gold treasures belonging to ancient Thracians dating back to as far as 5th millennium B.C. have been found in Varna Necropolis in Bulgaria. In 3100 B.C., the Egyptian ruler Menes laid the foundation for incorporating gold into the Egyptian economy and had decreed that "one part of gold in value was equal to 2.5 parts of silver". Gold was also an integral part of the Indus Valley civilization in India and was used in jewellery and as barter.

The first official declaration of gold as money came in 600 B.C. when King Alyattes of Lydia oversaw the first recorded mint. Coins which were an alloy of gold and silver called electrum were stamped with denominations. Darius I of Persia introduced a 95.83% pure 8.4g gold coin minted from his treasury, which was deemed equivalent to 20 silver coins.

Roman Society which had been using coins for exchange also introduced gold coins in 300 B.C. and this was continued by the Byzantine Empire until the middle ages. The Italian *Florin* became the most dominant gold coin along with the German *Augustalis* introduced under the Frederick II. By the 14th century England had also moved towards using gold as a currency by minting its coins known as *Noble*.

Unsurprisingly, most treasures in the world involved gold. One famous surviving treasure vault was discovered at the famed *Shree Padmanabhaswamy Temple* in Thiruvananthapuram, Kerala, India. The vaults of the temple are rumoured to hold an estimated Rs. 1 Trillion (US $20 Billion) of wealth in gold and other precious commodities in just one of its many vaults. The vault-B of the temple was opened and audited under the aegis of the Supreme Court of India and unimaginable wealth in the form of gold coins and necklaces were discovered inside.

Apart from being a part of treasures around the world, gold also helped to kick start the first banking system. The fractional banking system first took hold in England after goldsmiths started to re-lend gold that had been deposited with them for storage in lieu of receipts in the early gold markets. The confiscation of large amounts of gold as forced loan by King Charles I led to many merchants storing gold with the gold smiths. This was the genesis of the 'fractional banking system'. As a convenience and for the safety of the gold traded, the system would circulate the titles of ownership of gold rather than the gold itself.

In the year 1816, the UK officially defined the pound sterling relative to gold. With the entry of the United States of America, the 'classic' gold standard was adopted in 1879 and was further solidified in 1900. The Bretton woods agreement signed in 1944 laid the foundation for convertible currencies where countries settled their international balances in dollars and the US dollar became fully convertible to gold. The exchange rate applied at that time was $35/ounce which was the responsibility of the United States. While the major powers constantly undermined the gold standard it was not until 1971 that the Federal government in the United States jettisoned the 'gold standard' once and for all. From then on it was the dollar trade that became the central theme of global trade and the 'US dollar' became established as the standard and preferred currency for trade.

Wealth Redefined

"If money is your hope for independence you will never have it. The only real security that a man will have in this world is a reserve of knowledge, experience, and ability."

– Henry Ford

While 'cattle', 'land' and 'gold' were initial bedrocks of wealth. The industrial revolution created new forms of wealth. Industrialization brought on rapid growth, mass production of goods, large scale factories with a large workforce. These not only created jobs but also created multi-millionaires and billionaires who amassed unimaginable wealth.

A factory may only require small acreages but it had the capacity to churn out goods that got traded far and wide, making its owners unimaginably rich and powerful. The first few factories produced products including agricultural items, textiles, and machine parts. The invention of the assembly line famously pioneered by Ford, manufactured and assembled goods. For those days, the output production rates were phenomenal. Vast amounts of wealth were created in the process and wealth began to be redefined. One was wealthy not just by owning vast tracks of land and gold, ownership of companies that managed factories and supply chains and other such avenues, also created wealth.

Andrew Carnegie was the richest person in the world at a time a when wealth was amassed from production and selling of factory goods. He amassed his wealth namely from his vertically integrated iron and steel business where he owned the entire supply chain from the mines to the railroads and controlled the pricing of the end product.

Unlocking of Wealth and 'Paper Money'

The *'East India Company'* became the first publically traded company in the world. Trading across the globe was very risky as ships could get lost or get destroyed in storms, or have its load plundered by pirates and many a time it may suffer mutinies. These risks were too many for one investor to withstand and it was but natural that the 'East India Company' was formed to spread the risk and rewards of its endeavours. Thus it became the first limited liability company. It was formed in the year 1600 A.D and was called *'Governor and Company of Merchants of London trading with the East Indies'*. This helped to diversify risk and mitigate it, as its investments were spread over a number of voyages. This reduced risk and chance of catastrophic failure if investments in a particular voyage drew a blank. In the year 1602 A.D., the first shares of the 'East India Company' were released in the *Amsterdam stock exchange*.

While the *New York Stock Exchange (NYSE)* was formed in 1817 A.D., the *NASDAQ* was created in the year 1971 A.D. Since then, a number of stock exchanges have been created in a number of countries around the world. Large amounts of wealth today reside in these stock exchanges as paper money. A number of entrepreneurs have listed their businesses in the stock markets of the world and they dominate the Forbes list of richest people in the world.

The Oil Economy

The initial demand for oil was satisfied by hunting whales and extracting oil from its blubber by flensing. This became a major occupation across Europe, United States and many countries in East Asia. The Whaling industry reached its peak during the latter half of the 18th century. In fact, whale hunting was a significant part of the GDP of many countries during that time.

The discovery and production of oil from drilling and the ability to extract kerosene from coal, caused the whaling industry to wane. Until, the signing of 'The International Convention for the Regulation of Whaling', whale hunting continued. The convention not only brought on international action to prohibit and discourage the hunting of whales but also spoke of conservation of whale stock.

Ever since the industrial revolution, the demand for coal and oil has grown exponentially and businesses involved in the extraction of coal/oil, have and continue to mint a lot of money. Initially, the *'Carbon Economy'* was defined by coal mining. The demand for kerosene and gasoline has been growing ever since the automobile revolution. Right from the industrial revolution to the information revolution, the 'Carbon Economy' as it is called, has defined wealth in the 20th century right into the 21st century. The rise of personal transport and global trade with automobiles and gasoline/diesel powered transport has caused the oil economy to peak and with time find new peaks. In the 19th century, J.D.Rockefeller Sr. controlled 90% of the oil economy in the United States. At one point this made him the richest American until he was surpassed by Andrew Carnegie. Wealth lay rooted in the 'Carbon Economy' ever since.

The 'Carbon Economy' and the industrial dependence on fuel spurred the rise and growth of a number of countries especially those in the Middle-East.

These countries were endowed with vast quantities of natural wealth constituted by oil and gas. Due to the voracious demands of the 'Carbon Economy' their resources were much valued and were in great demand. Oil defined wealth and continues to do so. However, headwinds have begun to appear as there is talk that the 'Carbon Economy' would give way to 'Renewable Energy Economy' with the advent of Electric Vehicles and Solar power. Just as IC engines and thermal power plants defined the 20th Century, these new technologies that use renewable energy would define the 21st century.

The Computer and Chip Revolution

Talking about sources of wealth, the 20th Century has seen the rise of computers and faster chips. Sources of 'Intangible wealth' beyond just tangible products or traded commodities began to become significant. This has caused a paradigm shift in the way wealth is built and measured. Software codes were written to power the hardware, and its usefulness has led to its rise as a source of intangible wealth. Microsoft, a company that envisioned a PC on every desktop foresaw that its operating system would be running on every computer. It became one of the world's most sought after stocks in the late 20th century. Hardware companies such as Intel grew in tandem producing faster and faster microprocessors to power the ever increasing demand of software upgrades. Wealth in the late 20th century and even today lay in these intangibles. These wealth creating teeny-weenie chips are certainly far removed from wealth constituted by land and gold of yesteryears.

The Information Superhighway

The coming of the *'internet'* or the *'information superhighway'* simply began a revolution that spawned hundred others. The internet and flow of information provided the basis for smart entrepreneurs to reach out to markets, offer services, control supply chains, outsource manufacturing, market to consumers, take-in e-payments in a manner never seen before. A staggering number of businesses riding on the back of the internet were born and many continue to flourish even as new businesses with newer business models come into vogue. The information Superhighway provided the basis for 'Creative Destruction' at its best. Giants like 'Amazon' and 'Google' were born and have created new kinds of wealth. These companies have provided customers with tools to find exactly what they want and give them access to ways to fulfil their needs. There has been a tectonic shift as flow of information has increased efficiency of supply chains, opened new markets, increased convenience to customers, enabled targeted marketing and has even changed the way that people live their lives. This new wealth redefined the way money was seen since the late 20th Century.

The Mobile Revolution

The smart phone revolution is however the real game changer. Taking off from the feature phone, the smart phone is a powerful computing device in a customer's hand constantly relaying information to and fro to company servers via a variety of applications called 'Apps'. This has not only created new avenues to serve the customer but has made the computing device omnipresent and omnipotent. The customers can now hail a cab, lookup restaurants, order food delivery, shop, play games, pay and chat all through one device in his/her pocket.

There is a paradigm shift where customisation and in-depth customer information allowed one to cater more specifically and effectively to the customer's needs. New models for delivery of services and new models to reach out to each individual consumer, to get their attention, time and money, are effected by these technologies. These are new ways of reaching out and fulfilling business motives. This is the new reality, wealth has been created by taking a number of businesses right into a device in the hands of the consumer.

Social Media Revolution

It would be pertinent to mention the 'Social Media' revolution in this context. The 'Social Media' companies allow the consumer to use their services for free and in exchange they obtain the personal data of the consumers. These companies also own all posted data and comb them to understand the profile of the consumers and provide them with tailor made services. Better understanding of the customers would mean better targeted advertisements. Such services are invaluable to advertisers, product and service providers around the world. Rather than the 'spray and pray' advertising of the past, targeting occurs at a micro level and each of the customers can be individually and effectively marketed to. Any manufacturer or service provider could define his/her target market and these social media channels would be invaluable channel partners to reach out to the customers with their marketing messages. These manufacturing and service companies are willing to pay big dollars to anyone who helps them to sell more. That is how 'Social Media' giants rake in the *moolah*, sometimes cents at a time for every target that they connect with. Like little drops make an ocean, when we consider the extent of their global networks and number of users, these cents per click translates into billions of dollars in revenue.

Data the 'New Oil'

Consequently data has now become the 'new oil'. It has become the resource that companies seek to harness to be able to understand and to better serve their customers. Companies seem bent on gathering this data and deep mining it for insights to be able to sell better to their customer. Companies on social media like *Facebook, Linkedin, Whatsapp, Instagram, Twitter* etc, sit on a treasure trove of data that can be mined for a variety of new applications and services.

With *IoT (Internet of things)* and AI (Artificial Intelligence) taking center stage in the 21st century, we are at the helm of an age where large quantities of data will be available and exploited. In this regard the "raw material" is DATA. Companies are in a race to gather as much of this gold as possible. It significantly matters who owns the data and who has access to it. The entire wealth of the new age companies resides on 'server farms' in the form of this data.

AI requires huge amounts of data to feed bots and enable machine learning. This aids machine response to new situations as and when they arise. The Google car for instance has recorded over 1.5 million miles of testing providing invaluable insights that could help train the self-driving software to respond to the varied situations it may confront in the future. The learning is such that it may be able to respond to situations it may not have confronted in the past. These kinds of insights that help in training an AI system however need vast amounts of data. Another example is healthcare. As 'wearables' become a rage, and as more and more data is collected from the devices that people wear, greater insights can be obtained to make these devices more accurate and also to break new ground in the fitness and health segment. Hence it does not come as a surprise that bits of data on servers is the new wealth or as some call it, the 'new oil'.

As a civilization that measured wealth by literally counting sheep and livestock we have come a long way. Today bits and bytes of information are a major source of wealth and going forward as the world becomes richer it can be hoped that people may move towards a more holistic definition of wealth as defined at an individual level. Some call this the 'True Wealth'.

'True' Wealth

Philosophically speaking real wealth is more than just money. In the simplest terms *'True Wealth'* is the ability to live life on one's own terms. It's freedom to have and exercise choices.

Individually speaking there are different kinds of wealth.

You can be wealthy in...

- Money & Possessions (Financial Wealth)
- Fame & Fortune
- Health & Happiness (Health Wealth)
- Purpose & Direction in Life
- Family & Friends (Relationship Wealth)
- Love & Laughter
- Freedom & Time (Time Wealth)
- Talent & Wisdom (Wealth of Wisdom)
- Peace of Mind & Spirituality (Spiritual wealth)

As people move up the economic ladder, wealth as defined by money has diminishing rates of return and people look beyond it. People in many developed nations are valuing fewer working hours and more family time.

More and more people are beginning to believe that blind consumption with disregard to the environment would only hurt the planet and come back to bite us as a race. Movements to work and care for the environment are growing. People have begun to believe that environmental health is true wealth we could leave for the future generations. There is also greater awareness among the new generation to stay healthy. The shift towards vegetarianism and veganism is an example of how people are making better choices to not just safeguard the environment but also to take care of their health. Healthy environment and healthy living are maxims that people are swearing by today. They choose organic food and prefer sustainable agriculture. More and more people are taking their health and fitness seriously. This only goes to show that wealth has many dimensions. 'Health is wealth' goes an old adage and the new generation have made an early start. The world is moving towards holistic living where they value physical fitness and living a stress free life. In this context it would be pertinent to mention how popular *Yoga* has become.

Individual wealth seen through the prism of 'True Wealth' is a function of our environment, the food we consume, the water we drink and the air we breathe. These need to be healthy just like our immediate living environment. This is part of our 'True wealth'. This 'True Wealth' will be affected by our lifestyle and the choices we make. How we consume and how we discard what we don't need, all make up our *carbon foot print* and could build or break our future. As inheritors of the planet, the future generations are more attuned towards the need to save the environment. We have come a long way and we are on a path where our views and perceptions about wealth will change to include that which makes holistic and sustainable living possible.

As the world becomes wealthier, wealth will begin to be seen beyond just GDP and per capita income. The true physical, mental, spiritual health of the population will be seen as a measure of true wealth. GDH or *Gross domestic happiness* as a central measure will take hold once the world has been through this mindless and soulless materialism cycle. This will happen gradually with greater realisation. 'True wealth' seen holistically has begun to see the dawn. With greater realisation this century there is going to be a fundamental shift in what basically contributes to and constitutes wealth. The traditional definitions of wealth are going to undergo a major overhaul once again!

CONCLUSION

Where we will be in the future would be determined by how we absorb change into society and life right now

"The true civilization is where every man gives to every other, every right that he claims for himself."

— Robert Green Ingersoll

"What was fiction yesterday is a possibility tomorrow and a reality the day after. Welcome to the brave new world."

— Author

Technology is in a constant march forward bringing with it sweeping changes to society and culture and the way we live, think and breathe. We have already seen how the changes in technology have drastically changed our lives within a short span of a decade. The pace of this change has only increased and we will witness many more revolutionary breakthroughs in life which would disrupt the old and give way to the new.

Humans continue to strive to push the frontiers of knowledge and are breaking new ground with each passing day. There is a theory akin to Moore's law that states that humans double their knowledge and what they know every two years. The compounding effect of this is sure to be staggering to say the least. The future sure looks ripe with possibilities. What was fiction yesterday are possibilities tomorrow and a reality the day after. Welcome to the brave new world of change and more change.

Conclusion

The 7 Deadly Sins

"The moment you give up your principles, and your values, you are dead, your culture is dead, your civilization is dead. Period."

– Oriana Fallaci

"Without memory, there is no culture. Without memory, there would be no civilization, no society, no future."

– Elie Wiesel

The more things change, the more they remain the same. In this constant flux it would help to recall what the great Mahatma said. Mahatma Gandhi helps us to center and bring us back to the core values that would help humanity survive and not self-destruct. He brings us to our base line, reminding us of the need to strike a balance lest we crash and burn. I would like to end this book with the seven deadly sins he says society should be wary off. In his words

"The 7 Deadly sins are

- Wealth without work,
- Pleasure without conscience,
- Science without humanity,
- Knowledge without character,
- Politics without principle,
- Commerce without morality,
- Worship without sacrifice."

While these concepts may seem abstract, it is necessary for us to take a step back and ponder philosophically about human civilization and where we are taking it. In the face of the barrage of technology and change we should not be swept so far away from our moorings that we begin to drift without direction and purpose. Hence no matter how much we progress on the scientific front, our society should NOT shed its core principles so that human beings survive together as a species. I leave you with these thoughts as I sign off.

Thanks for reading,

A. Venkatasubramanian

END NOTE

While it may not be possible to anticipate all disruptions led by technologies across multiple industries, this book has brought out key forerunners which have direct and larger impact on the underlying economy, society and systems around it. The understanding of this along with a creative twist on future markets, k-commerce and other interesting snippets should ignite the minds of new-age entrepreneurs.

These entrepreneurs who are backed with serial investors are the torch bearers of change who support these constant breakthroughs over a period of time. This is why it is necessary for the entrepreneurs and investors to keep abreast of all technological changes and tools available to them.

As entrepreneurs and investors are change agents with very clear goals, their passion should be channelled towards what is not only profitable as a business venture but also in what is beneficial to society and the environment.

– Dheeraj Jain,

MD at Redcliffe Capital

ABOUT THE AUTHOR

Mr. A.Venkatasubramanian is the author of 'You Better Watch Out!' and has authored three other books before this one. He has also penned several articles that were published online and featured in newspaper columns.

Mr. A.Venkatasubramanian is a **qualified engineer** and **management graduate** having done his **B.E.** from INDIA and **M.S.** in Engineering and **M.B.A** in General Management and Finance from the US. He has also completed **CFA** (Chartered Financial Analyst, USA) until Level - II.

About the Author

While he **started out his career in the technology industry,** his experience spans a number of industries including **automobile, airline, information technology, retail, banking, real estate, market research** and **finance.**

Mr. A. Venkatasubramanian is currently an **entrepreneur in the technology space** and loves **travel, photography, music** and **dance.** He is an **avid reader** and **sportsman.**

Twitter Handle	Facebook
@AVenkatasubram4	facebook.com/A.Venkat.Author/

Also Read

A never before insight and approach to investment in REAL ESTATE

www.therealdealbook.net
www.redbrickgreenback.com

Also Read

Wisdom on investments presented as 'Articles' and 'Anecdotes' which will leave the reader enriched, wiser and wealthier.

www.therealdealbook.net
www.redbrickgreenback.com

Also Read

A Book on Investments where
Western Materialism meets Eastern Philosophy

www.redbrickgreenback.com
www.therealdealbook.net

Also Read

Official Website

The Official Website has been designed to promote the Trilogy of books on Investments
www.TheRealDealBook.net

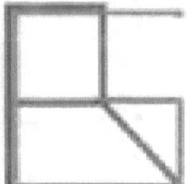

Red Brick Green Back Consulting

Red Brick Green Back Consulting has set out on a mission to educate people about investments in 'Real Estate' and 'Financial Instruments'

Website

www.RedBrickGreenBack.com

Also Read

How 'Technologies of the Future' are going to change and impact Society, Culture and YOUR LIFE

www.ingramcontent.com/pod-product-compliance
Lightning Source LLC
Chambersburg PA
CBHW030929180526
45163CB00002B/503